'U S
Y
4

the
jes
s.

UNIVERSITY
COLLEGE CHESTER
WARRINGTON CAMPUS

essential quantiative methods

*To Doreen, Jim and Family*

A GUIDE FOR BUSINESS

# DONALD WATERS

essential quantitative methods

FINANCIAL TIMES
Prentice Hall

*An imprint of* **Pearson Education**

Harlow, England · London · New York · Reading, Massachusetts · San Francisco · Toronto · Don Mills, Ontario · Sydney
Tokyo · Singapore · Hong Kong · Seoul · Taipei · Cape Town · Madrid · Mexico City · Amsterdam · Munich · Paris · Milan

**Pearson Education Limited**
Edinburgh Gate
Harlow
Essex CM20 2JE
England

and Associated Companies throughout the world

*Visit us on the World Wide Web at:*
http://www.pearsoned.co.uk

Typeset by 32
Typeset in 10/12 Times
Produced by Pearson Education Asia (Pte) Ltd.
Printed in Singapore

First printed 1998

ISBN 0–201–33137–3

**British Library Cataloguing-in-Publication Data**
A catalogue record for this book is available from the British Library

**Library of Congress Cataloging-in-Publication Data** is available

10  9  8  7  6  5  4  3
07  06  05  04  03

# Contents

## Appendices

# Preface

Quantitative methods are widely used in business. This book brings together the core material in a concise form. Its main features are:

- it is an introductory text and assumes no previous knowledge of management or quantitative methods;

- it can be used by students for a wide range of business courses, or by individuals studying by themselves;

- it concentrates on core material, and gives a concise description of the most important material;

- it describes methods that have proved useful in practice;

- it illustrates principles by examples rather than by theoretical discussion;

- it omits details of mathematical proofs and derivations;

- it uses computer output – particularly spreadsheets – to illustrate calculations;

- it has a friendly style with many learning features.

## Audience

A growing number of students are taking courses in management. Some of the courses give a broad description of business, while others specialise in accountancy, marketing, human resources, or some other function. The courses are run by colleges, universities, professional organisations and companies. Despite the diversity of courses, most business students attend a course on Quantitative Methods.

Managers need to understand some quantitative methods – but they do not have to be mathematicians. This book is aimed directly at managers. It takes a friendly look at essential material, and the contents are practical rather than theoretical. A deliberate decision has been made to avoid rigorous (and often tedious) mathematics and concentrate on applications rather than theory. Many of the calculations are illustrated by computer printouts – often using spreadsheets.

# Format

Each chapter has a consistent format which comprises:

- a list of contents for the chapter;

- a list of things a reader should be able to do by the end of the chapter;

- the main material of the chapter divided into pertinent sections;

- worked examples to illustrate methods described;

- a summary of the main points at the end of each section;

- a review at the end of each chapter listing the material that has been covered;

- additional problems;

- a case study.

Most people have access to a computer so there is no point in doing calculations by hand. Nevertheless, the book does not assume that readers have access to a computer, and does not use specific packages.

# Contents

Almost any topic in mathematics might be useful to managers in some circumstances, so there is a wide range of material that could be put into a book of this type. In order to keep the book to a reasonable length, only the most widely used methods have been included. Thankfully, there is increasing agreement about which methods these are. The book covers the core material that is generally included in quantitative methods courses for managers.

The book is divided into four parts which develop the subject in a logical sequence. Part One gives an introduction to the subject; Part Two covers the collection and description of data; Part Three shows how these ideas are used for different business problems; Part Four describes some useful statistics. Students often find probabilistic ideas much more difficult that deterministic ones, hence there is a clear separation of deterministic methods (described in Parts Two and Three) and probabilistic methods (described in Part Four).

All in all, the book provides students with a solid foundation for understanding quantitative methods and their use in business.

part

1

# Introducing quantitative methods

Part One introduces some ideas of quantitative methods and lays the foundations for the rest of the book.

There are two chapters in Part One. The first chapter gives a general introduction to quantitative methods, while the second describes some arithmetic that is used in the rest of the book.

# 1 Numbers and managers

## Contents

## Chapter outline

This chapter gives a general introduction to quantitative methods. After reading the chapter you should be able to:

- understand the importance of numbers
- say why organisations use quantitative methods
- understand the use of models
- describe a general approach to solving problems

# 1.1 Who needs numbers?

We are surrounded by numbers. On a typical day you might find that the temperature is 17°C, petrol costs 69p a litre, 2.1 million people are unemployed, a company reports a profit of £17 million, gold costs $317 an ounce, a cricket team scores 274 runs, and 82 per cent of people want more shops to open on Sunday.

We can only live a normal life if we know how to use numbers. If you have any doubts about this, imagine living without numbers. You could not buy anything because you would not understand the price; you could not travel anywhere because you would not know how far to go; you could not follow sports because the scores would have no meaning; you could not describe how old you are, or how much you weigh, or how much you earn, or how fast you can run, or how big your house is, or who won an election.

We use numbers to describe things. They give an objective measure that is both accurate and precise. A house has four bedrooms, the next town is ten miles away, 187 people watched a swimming match, ICI shares cost £7.47 each, a marathon is 26.2 miles long, families have an average of 1.8 children. When you cannot use numbers to measure something, it is much more difficult to understand. If you go to the doctor with a pain in your stomach, it is difficult to describe accurately and concisely what kind of pain you have, how bad it is or how it makes you feel.

We also use numbers in calculations. If bars of chocolate cost 30 p each and you buy three bars, the total cost is $3 \times 30 = 90$p. If you pay for these with a £5 note you will get £4.10 in change.

Sometimes we have to do calculations accurately. If car mechanics charge £10.50 an hour and they work on a car for two and a half hours, they will charge $10.5 \times 2.5 = £26.25$ for their time. Often, though we only need a rough estimate of the results. If you are planning a 240 mile journey and travel at 50 miles an hour, you know the journey will take around five hours. If you see apples costing 78p a kilogramme, you know that five kilogrammes will cost about £4.

We all use numbers in our everyday life, but we do not have to be expert mathematicians to understand quantitative ideas. Worked Example 1.1, shows how a simple calculation demonstrates the best value for money.

## we Worked example 1.1

A jukebox only accepts 10p coins. The number of plays given are:

10p – 1 play

20p – 3 plays

30p – 4 plays

40p – 5 plays

What is the best way to use the machine?

## Solution

Customers get the best value when the cost for each play is lowest.

- Paying 10p gives 1 play, so the cost is $10/1 = 10.0$p a play.
- Paying 20p gives 3 plays, so the cost is $20/3 = 6.7$p a play.
- Paying 30p gives 4 plays, so the cost is $30/4 = 7.5$p a play.
- Paying 40p gives 5 plays, so the cost is $40/5 = 8.0$p a play.

As you can see, paying 20p gives the lowest cost per play.

The main problem with arithmetic is that we all make mistakes with even the simplest calculation. Thankfully, computers now do the calculations for us, but we still have to understand what the computer is doing, and what the results mean. Spreadsheets like Excel, Lotus 123 and QuattroPro are particularly useful. Figure 1.1 shows a spreadsheet for the calculations in Worked Example 1.1. In this, you can see the costs in column A, and the number of plays this buys in column B. Column C shows the cost per play, which is column B divided by column A. Cell C9 shows the lowest value in column C, and this is marked by stars in column D. These calculations are shown in Figure 1.2.

|   | A | B | C | D |
|---|---|---|---|---|
| 1 | Cost of using jukebox | | | |
| 2 | | | | |
| 3 | Cost | Number of plays | Cost per play | Best option |
| 4 | | | | |
| 5 | 10 | 1 | 10.00 | |
| 6 | 20 | 3 | 6.67 | ***** |
| 7 | 30 | 4 | 7.50 | |
| 8 | 40 | 5 | 8.00 | |
| 9 | Lowest | | 6.67 | |

**Figure 1.1** Finding the best option on a jukebox.

|   | A | B | C | D |
|---|---|---|---|---|
| 1 | Cost of using jukebox | | | |
| 2 | | | | |
| 3 | Cost | Number of plays | Cost per play | Best option |
| 4 | | | | |
| 5 | 10 | 1 | =+A5/B5 | =IF(C5=$C$9,"*****"," ") |
| 6 | 20 | 3 | =+A6/B6 | =IF(C6=$C$9,"*****"," ") |
| 7 | 30 | 4 | =+A7/B7 | =IF(C7=$C$9,"*****"," ") |
| 8 | 40 | 5 | =+A8/B8 | =IF(C8=$C$9,"*****"," ") |
| 9 | Lowest | | =MIN(C5:C8) | |

**Figure 1.2** Spreadsheet calculation to get Figure 1.1.

In this book we will assume that computers do all the difficult calculations. We will describe the analyses, but remember that:

## Computers do arithmetic

Quantitative methods are particularly important in business. Managers often talk about 'improving productivity', 'increasing return on investment', 'scheduling production', 'forecasting demand', 'increasing customer service', and so on. All of these rely on measurements and calculations. An organisation's overall performance is summarised in its accounts – which are largely numerical. You can see, then, that quantitative ideas are at the heart of every organisation.

We are going to show how quantitative methods are used in business. Basically, they give managers information and make suggestions. Then managers look at all the information they have – which is both quantitative and qualitative – and use their skills, knowledge and experience to make decisions, as the flowchart in Figure 1.3 indicates.

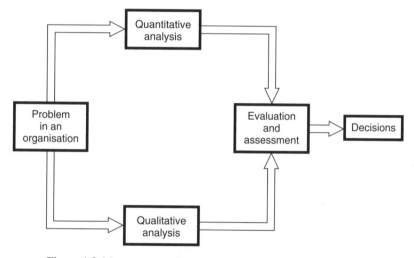

**Figure 1.3** Managers use all available information to make decisions.

Remember that the analyses we describe give information but:

## Managers make decisions

## we Worked example 1.2

Synogistic Management Consultants aim to have 10 clients on its books for every consultant that they employ. Last month they had 125 clients on their books. How many consultants should they employ?

## Solution

If Synogistic takes a purely quantitative view, it would employ 125/10 = 12.5 consultants. This figure could be rounded to either 12 or 13, or a part-time consultant could be employed. However, before making any decision, Synogistic has to look at qualitative factors, like the expected changes in business, attitudes of consultants, type of business, planned staff departures and arrivals, morale, and competition. When the company has considered all the available information, it can make its final decision.

### In summary

**Numbers are an essential part of everyday life. We can use them to describe things, and for calculations. They are particularly important in business – so managers must understand some quantitative methods.**

# 1.2 Using models

This book shows how to solve different types of problems. The first step is always to build a **model** of the problem. You probably think of models as toys – but here we are using them to describe real situations. The main characteristics of our models are:

- they describe real situations;
- they are simplified, and only include relevant details;
- properties in the real world are represented by other properties in the model.

**Models are simplified descriptions of real situations**

Suppose a company makes a product for £200 a unit and sells it for £300 a unit. Then we can describe the company's profit by the equation:

profit = number of units made $\times$ (selling price $-$ cost)

or     $P = N(300 - 200) = N \times 100$

This equation is a simple model, with $P$ representing the profit and $N$ the number of units made. We can do experiments on this model – to see how the profit changes with different production levels. For this, we just take the equation $P = N \times 100$, substitute different values for $N$ and find the corresponding values for $P$.

Of course, we could experiment with the actual operations, by changing the production and measuring the profits, but this is time consuming, difficult, expensive and could seriously damage the company. It is easier, cheaper, faster and less risky to experiment with a model than to experiment with real operations. Indeed sometimes it is impossible to experiment with real operations. For example, a company cannot find the best location for a factory by trying every possible location and keeping the best. Nor can it choose a new product by starting to make all possible new products and then scrapping those it does not like. So, a model gives the best – and often the only – way of analysing a situation.

## In summary

**Quantitative methods are based on models. These are simplified views of real situations. Most models in business use equations, with real things represented by symbols.**

# 1.3  How to tackle a problem

Many different quantitative models are used in business. Some of these are very simple and easy to understand; others are complicated and take years to develop. Some operations are so complicated that nobody has yet built a reasonable model; and some models are so complicated that no-one has been able to solve them. Despite the differences in models, we can use a standard approach for solving business problems. This approach has four stages:

(1) **Observation**, where managers examine the problem, collect data, describe the details of the problem, set objectives, consider the context of the problem and discuss different ideas.

(2) **Modelling**, where managers analyse the data, build a model, test it, find initial solutions and check these to make sure they are realistic.

(3) **Experimentation**, where managers examine the initial solutions, search for better answers, consider alternative values for variables, collect any other data needed, and make recommendations.

(4) **Implementation**, where managers make the final decisions, set values for variables, implement these decisions, monitor actual performance and update the model when necessary.

The length and difficulty of these stages depend on the problem tackled. An oil company, like Shell or BP, can spend years finding its best production policies. A village shop can come to a minor decision in a few minutes.

## we  Worked example 1.3

John Fraser opened an electrical shop in Brighton in 1990. Business was good, and in 1996 when the shop next door became vacant, he considered an expansion. How do you think he tackled this decision?

# Solution

John tackled the decision in four stages.

(1) First, he looked at the options. These included expanding into the shop next door, opening a second shop somewhere else, closing his existing shop and moving somewhere else, and staying with his existing shop. He collected information about the costs of the shop next door and other locations. He also talked to different people about grants, trade prospects, local traffic and competition. This was the observation stage.

(2) Then John combined the information into a series of cost and profit forecasts for different options. He checked the results against the actual performance of his own shop over the past year, against published figures for other shops nearby, and against figures from his competitors. He had to adjust his initial models several times until he was convinced that his forecasts were accurate. This was the modelling stage.

(3) Then John adjusted his models to see what would happen in different circumstances. For example, he found how his profit would change if rents rose faster in Brighton than in other towns. This was the experimentation stage.

(4) Finally, he looked at the results and decided to open a second shop in another location rather than expand next door. It took eight months to organise the expansion, but John now has a second shop running successfully in Eastbourne. This was the implementation stage.

Business decisions are not made in isolation, but are part of a continuing management process. Each new decision must fit-in with other, related decisions. So managers look at the effects of earlier decisions, before making any new decisions. This is the basis of feedback, shown in Figure 1.4 opposite. Here a manager uses a model to get results, make decisions and implement them. Then feedback returns the effects of the decisions to the manager, who can use them to update the model.

## In summary

**There are many types of model, but we can use a standard approach to solving problems. This has four stages: observation, modelling, experimentation and implementation.**

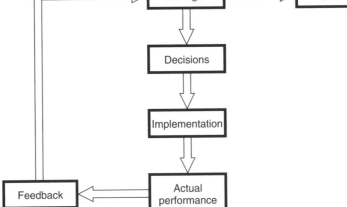

**Figure 1.4** Feedback in decision making.

# Chapter review

This chapter introduced the ideas of quantitative methods. In particular it described:

- how we are surrounded by numbers – which we use to describe things and in calculations;
- how quantitative analyses are particularly important in business;

- the use of models;

- an approach to solving business problems, based on observation, modelling, experimentation and implementation.

# Problems

1.1    In last year's Southern Florida Amateur Tennis Championships there were 1947 entries in the women's singles. If this was a standard knock-out tournament, how many matches were played to find the champion?

1.2    What is the smallest number of coins needed to pay exactly a bill of £127.87?

1.3    A family of three is having grilled steak for dinner. They like to grill their steaks for ten minutes on each side. Unfortunately, their grill pan is only big enough to grill one side of two steaks at a time. How long will it take to cook dinner?

1.4    A shopkeeper buys a vase for £25. A customer buys it for £35 and pays with a £50 note. The shopkeeper does not have enough change, so he goes to a neighbour and changes the £50 note. A week later he is told that the £50 note was a forgery, so he immediately repays his neighbour the £50. How much does the shopkeeper lose?

1.5    See what computers you have available. Find out what programs there are, including word processing, spreadsheets, databases, graphics and statistics packages. Practise using these programs.

1.6    Table 1.1 shows the number of units of a product sold each month by a shop, the amount the shop paid for each unit, and the selling price. Use a spreadsheet to find the total values of sales, costs, income and profit.

**Table 1.1**

| Month | Units sold | Cost to the shop | Selling price |
|---|---|---|---|
| January | 56 | 120 | 135 |
| February | 58 | 122 | 138 |
| March | 55 | 121 | 145 |
| April | 60 | 117 | 145 |
| May | 54 | 110 | 140 |
| June | 62 | 106 | 135 |
| July | 70 | 98 | 130 |
| August | 72 | 110 | 132 |
| September | 43 | 119 | 149 |
| October | 36 | 127 | 155 |
| November | 21 | 133 | 161 |
| December | 22 | 130 | 161 |

# CS Case study – Hamerson Builders' Merchants

Albert Hamerson is managing director of his family firm of builders' merchants. He is the third generation to run the company, and his daughter, Georgina, is keen to join him when she leaves university. However, she is not sure about the kind of job she wants.

Hamerson is a wholesaler. The company buys 9000 different products from 1200 manufacturers and importers, and sells everything that builders need. It works from five sites around Liverpool and employs over 300 people.

Georgina feels that the company is getting behind the times. When she goes into the office she is surprised at the amount of paperwork, as she had assumed that computers were leading to a paperless office. When she walks around the stores, things still seem to be organised in the way they were twenty years ago. Georgina has some ideas for improving things, and wants the chance to develop these. She imagines herself as an internal consultant looking around the company, finding areas that could be improved, and then doing projects to make operations more efficient.

The problem is that Georgina is doing a course in mathematics and business studies. Apart from reading the accountants' reports, Albert has never had any contact with quantitative ideas. After a lot of discussion, they agree that Georgina should write a report to show how the company can use her mathematical training.

If you were Georgina, what would you write in your report?

# 2 Tools for quantitative methods

## Contents

## Chapter outline

This chapter describes some basic mathmatics and algebra. We will use these in later chapters, so it is important that you understand them. After reading the chapter you should be able to:

- do arithmetic using integers, fractions, decimals and percentages
- use of algebra to describe problems
- solve equations with one unknown
- solve simultaneous equations
- use powers and scientific notation

# 2.1 Working with numbers

Before we look at some specific models that are used by managers, we have to revise some basic mathematics. But remember that we are looking at the *use* of quantitative methods, so we are taking a *practical approach*, and will not get bogged down in the details of proofs, derivations or arithmetic. We will introduce ideas by examples – and will always use computers for calculations.

We must assume that you are familiar with numbers and arithmetic. You can do some calculations, and see that:

- if you buy 10 loaves of bread at 72p a loaf, you will pay £7.20;

- if you drive at 80 kilometres an hour, you can complete a 400 kilometre journey in 5 hours;

- if you spend £300 a month on your house, £100 a month on food, and £200 a month on other things, your expenses are £600 a month; this is the same as $600 \times 12 = £7200$ a year, or $7200 \div 52 = £138.46$ a week;

- if a company has an income of £1 million a year and costs of £0.8 million a year, it makes a profit of £0.2 million or £200 000 a year.

There are four basic operations in arithmetic: addition, subtraction, multiplication and division.

> **Arithmetic operations**
>
> $+$   addition      e.g. $2 + 7 = 9$
> $-$   subtraction    e.g. $15 - 7 = 8$
> $\times$   multiplication   e.g. $4 \times 5 = 20$
> $/$   division       e.g. $12/4 = 3$

There are other ways of writing these operations:

- division can be written as $12 \div 4 = 3$, or $\frac{12}{4} = 3$, or $12/4 = 3$

- multiplication can be written as $3 \times 2 = 6$, or $3.2 = 6$ or $3*2 = 6$

Computers use the standard notation $12/4 = 3$ and $3*2 = 6$.

Sometimes you have to be careful with calculations. When you see $3 + 4 \times 5$, do you do the addition first and get $7 \times 5 = 35$, or do you do the multiplication first and get $3 + 20 = 23$? The general rule is to multiply before adding, so the second of these is right. If there is any doubt, you can use brackets to make the meaning clear. Then calculations in brackets are done first, so:

$$(3 + 4) \times 5 = 7 \times 5 = 35$$

and

$$3 + (4 \times 5) = 3 + 20 = 23$$

If one set of brackets is not enough you can use more, with some brackets inside others. Then you start calculations from the innermost set of brackets and work outwards.

So:    $4 \times ((5 - 3) \times 4)/4 - 2 = (4 \times 2 \times 4)/4 - 2 = 8 - 2 = 6$

while:    $(4 \times 5) - ((3 \times 4)/(4 - 2)) = 20 - (12/2) = 20 - 6 = 14$

The order of calculations is always the same:

- Calculations inside brackets – working from inner sets outwards.
- Raising to powers (which we mention later in the chapter).
- Multiplication and division in the order that they appear.
- Addition and subtraction in the order that they appear.

## we Worked example 2.1

What are the values of:

(a)  $(120/6)$

(b)  $(-120)/(-6)$

(c)  $\{(-2 \times 4) \times (15 - 17)\} \times (-2)$

(d)  $(((10 + 20) - (3 \times 7)) + (16/4))$

## Solution

Using the standard rules gives:

(a)  $(120/6) = 20$

(b)  $(-120)/(-6) = 20$   (Notice that the minus signs cancel each other)

Remember that multiplying two negative numbers together gives a positive number, while multiplying a positive number by a negative number gives a negative number.

(c)  $((-2 \times 4) \times (15 - 17)) \times (-2) = ((-8) \times (-2)) \times (-2) = 16 \times (-2) = -32$

(d)  $(((10 + 20) - (3 \times 7)) + (16/4)) = ((30 - 21) + 4) = 9 + 4 = 13$

The numbers in Worked Example 2.1 are integers. In other words they are whole numbers, like 20, 9 or 150. Integers can be positive, such as 3, 100, 257; or they can be negative, like $-2$, $-157$, $-356$. To make long numbers look better, you can use spaces or commas to divide the digits into groups of three, giving 1 247 822 or 1,247,822.

Often we have to divide integers into smaller parts. When two people share a bar of chocolate they get a half each. We can describe these parts of integers by **common** or **vulgar fractions** (like 1/2 or 1/4) or **decimal fractions** (like 0.5 or 0.25). In practice, 'common fraction' is usually abbreviated to just 'fraction'.

## we Worked example 2.2

Find the value of the following as decimal fractions:

(a) $\dfrac{36}{8}$

(b) $\dfrac{1}{2} + \dfrac{4}{5}$

(c) $\dfrac{3}{4} - \dfrac{1}{6}$

(d) $\dfrac{1}{4} \times \dfrac{2}{3}$

(e) $\left(\dfrac{3}{5}\right) \div \left(\dfrac{4}{5}\right)$

(f) $\dfrac{3}{6} \times \dfrac{2}{5} \div \dfrac{3}{4}$

## Solution

You can change common fractions to decimal fractions by straightforward division (preferably using a calculator). Then:

(a) $\dfrac{36}{8} = 4.5$

To add or subtract fractions, all numbers below the lines (called the **denominators**) are made the same, and then numbers above the lines (called the **numerators**) are added or subtracted.

(b) $\dfrac{1}{2} + \dfrac{4}{5} = \dfrac{5}{10} + \dfrac{8}{10} = \dfrac{13}{10} = 1.3$

(c) $\dfrac{3}{4} - \dfrac{1}{6} = \dfrac{9}{12} - \dfrac{2}{12} = \dfrac{7}{12} = 0.583$

To multiply fractions, all the numerators are multiplied together, and all the denominators are multiplied together.

(d) $\dfrac{1}{4} \times \dfrac{2}{3} = \dfrac{1 \times 2}{4 \times 3} = \dfrac{2}{12} = \dfrac{1}{6} = 0.167$

For division of fractions, you turn the fraction which is dividing upside-down and then multiply by it.

(e) $\dfrac{3}{5} \div \dfrac{4}{5} = \dfrac{3}{5} \times \dfrac{5}{4} = \dfrac{3 \times 5}{5 \times 4} = \dfrac{15}{20} = 0.75$

(f) $\dfrac{3}{6} \times \dfrac{2}{5} \div \dfrac{3}{4} = \dfrac{3}{6} \times \dfrac{2}{5} \times \dfrac{4}{3} = \dfrac{3 \times 2 \times 4}{6 \times 5 \times 3} = \dfrac{24}{90} = 0.267$

## we Worked example 2.3

A Canadian visitor to Britain wants to change $250 into pounds sterling. The exchange rate is $2.15 to the pound and banks charge a fee of £5 for the conversion. How many pounds does the visitor get?

## Solution

$250 is equivalent to 250/2.15 = £116.28. The bank then takes its fee of £5 to give the visitor £111.28.

Another way of describing fractions is to use **percentages**. Percentages are fractions where the bottom line is 100, and the '/100' has been replaced by the abbreviation '%'. Then if you hear, '20% of the electorate did not vote in the last election' you know that 20 people out of each 100 did not vote. You could describe this as a:

- common fraction:   $\frac{20}{100} = \frac{1}{5}$
- decimal fraction:   0.2
- percentage:   20%

If you hear that a company made a profit of £12 million last year and £3 million of this came from exports, you can say that:

- the fraction of profit from exports = 3/12 = 1/4
- the decimal fraction of profit from exports = 0.25
- the percentage of profit from exports = 25%

## we Worked example 2.4

What is:

(a)  17/20 as a percentage?

(b)  80% as a fraction?

(c)  35% as a decimal fraction?

(d)  40% of 120?

## Solution

(a)  $17/20 = 85/100 = 85\%$

(b)  $80\% = 80/100 = 4/5$

(c)  $35\% = 35/100 = 0.35$

(d)  40% of $120 = 0.4 \times 120 = 48$

## we Worked example 2.5

A shop sells dining room tables for £400 each. At one time it was difficult to get supplies of the table, so the shop raised the price by 20%. When supplies returned to normal the shop reduced the higher price by 20%. What was the final selling price?

## Solution

The normal price was £400. During the shortage the shop raised this by 20% to 120% of £400. This is:

$$\frac{120}{100} \times 400 = 1.2 \times 400 = £480$$

At the end of the shortage the shop reduced the higher price by 20%, to give a final selling price of 80% of £480. This is:

$$\frac{80}{100} \times 480 = 0.8 \times 480 = £384$$

If you take a fraction like $\frac{2}{3}$, this is equivalent to 66.67%, but the exact answer is 66.666 666 6...%, where the dots mean there is an unending row of sixes. To get 66.67 we have rounded this number to:

- two decimal places, showing only two digits after the decimal point, and

- four significant figures, showing only the four most important digits.

We round numbers to make sure that we get enough information but are not swamped by too much detail. There is no general rule for the best number of decimal places or significant figures, but we can give two suggestions.

- Only give the number of figures that is useful. Then you should avoid figures like £120.347 826 59 and quote this as £120.35, or even £120.

- Results from calculation are only as accurate as the data used. If you multiply a forecast demand of 32.63 units by a unit cost of £17.19 you should not describe the forecast cost as $32.63 \times 17.19 = £560.9097$. The original data have two decimal places, so the answer should not give more than this – or £560.91. In practice, this might be rounded to £561, £560 or £600 depending on how the figures are to be used.

Remember, when you are rounding to, say, two decimal places and the digit in the third decimal place is 0, 1, 2, 3 or 4 you round **down**, and when it is 5, 6, 7, 8 or 9 you round **up**. Then 11.364 is 11.36 to two decimal places, and 11.4 to one decimal place. You should only do this rounding on the final result, to reduce the effect of 'rounding errors'.

## we Worked example 2.6

What is 21 374.341 481 2 to:

(a)  two decimal places

(b)  four decimal places

(c)  two significant figures

(d)  four significant figures?

## Solution

(a)  21 374.341 481 2 is 21 374.34 when rounded to two decimal places.

(b)  21 374.3415 when rounded to four decimal places.

(c)  21 000 when rounded to two significant figures.

(d)  21 370 when rounded to four significant figures.

## In summary

**There are standard rules for doing calculations. You can use these rules for calculations with integers, decimals, fractions and percentages. Then you can round the results to a reasonable number of decimal places or significant figures.**

# 2.2  Changing numbers to letters

## 2.2.1  Forming equations

The cost of running a car depends on the distance that you travel. In one month you might drive 6000 km at a cost of £2400. Then you could say:

$$\text{cost per kilometre} = \frac{\text{total cost}}{\text{number of kilometres travelled}} = \frac{£2400}{6000} = £0.40 \text{ per km}$$

Rather than writing the equation in full, we could save time by using abbreviations. We could abbreviate the total cost to $T$, which is simple and easy to remember. We could also use $C$ for cost per kilometre and $K$ for number of kilometres travelled. Then our equation becomes:

$$C = \frac{T}{K}$$

The only difference between this and the original equation is that we have used shorter names to describe the variables. This is the basis of **algebra**.

We chose the abbreviations $C$, $T$ and $K$, as they were easy to remember, but we could have chosen other names, to give, for example:

$$c = \frac{t}{k}$$

$$y = \frac{x}{z}$$

$$\text{COST} = \frac{\text{TOTAL}}{\text{KILOM}}$$

or

$$\text{COSTPERKM} = \frac{\text{TOTALCOST}}{\text{KILOMETRES}}$$

Provided that the meaning is clear, it does not matter what abbreviations you use. But sometimes you have to be careful. The multiplication sign between variables is often assumed, so that $2 \times a$ is written as $2a$, and $4 \times a \times b \times c$ is written as $4abc$. This works if you use single-letter abbreviations, but can be misleading with longer names. If you abbreviate the total unit cost to TUC, it makes no sense to write an equation as

$$\text{TUC} = \text{NTUC}$$

when you really mean

$$\text{TUC} = \text{N} \times \text{T} \times \text{UC}$$

Obviously, you have to use common sense with your choice of names, and put in the arithmetic operators explicitly when there might otherwise be misunderstanding.

Returning to our equation $C = T/K$, we can rearrange this by multiplying both sides of the equation by $K$ to get:

$$C \times K = \frac{T \times K}{K} \qquad \text{or} \qquad T = C \times K$$

Now, if we divide both sides of this equation by $C$, we get:

$$\frac{T}{C} = \frac{C \times K}{C} \qquad \text{or} \qquad K = \frac{T}{C}$$

So we can rearrange an equation to give different views of the same situation. The equation remains true, provided we do the same thing to both sides. So we can multiply both sides by 2 to get:

$$2C = \frac{2T}{K}$$

and add 10 to get:

$$2C + 10 = \frac{2T}{K} + 10$$

and divide by $3A$, where $A$ is constant, to give:

$$\frac{2C + 10}{3A} = \frac{2(T/K) + 10}{3A}$$

**Provided you do the same thing to both sides of an equation, the equation will remain true.**

## In summary

**You can use abbreviations to identify values. This is the basis of algebra. Algebra gives a concise way of describing situations. Calculations in algebra are done in exactly the same way as calculations with numbers.**

## 2.2.2 Solving simple equations

In our equation for the cost of running a car, $C = T/K$, the cost per kilometre is fixed and we cannot change it. In other words, $C$ is a constant. However we can change the number of kilometres we drive and the resulting cost. So we have:

- **constants** which have fixed values;

- **variables** which are not fixed and can change values.

Equations show the relationship between constants and variables.

If we know the values of some constants and variables, and we have an equation relating them, we can find the value of a previously unknown constant or variable. This is called **solving an equation**.

The easiest way of solving an equation is to arrange it so that the unknown value is on one side of the equals sign and all the known values are on the other side. Suppose the managers of a company spend an average of $H$ hours a week attending meetings, and their time is worth $C$ an hour. The total annual cost of attending meetings, $T$, is:

$$T = 52 \times H \times C$$

If we know the values of $H$ and $C$, we can solve the equation to find the unknown value of $T$. If the managers spend an average of 15.2 hours a week attending meetings, and their time is worth £22.40 an hour, we can solve the equation to find:

$$T = 52 \times H \times C = 52 \times 15.2 \times 22.40$$
$$= £17\,704.96 \text{ a year}$$

Remember, though, that results from equations cannot be more accurate than the data used. We only know $H$ and $C$ to three significant figures, so we can only really give the answer to three significant figures – which is £17 700.

If we have a single equation we can solve this to find the value for **one** unknown. If we have several unknown variables, we need more equations – as we shall see in the following sections.

## we Worked example 2.7

A company employs 10 people and has total costs of £250 000 a year. These costs include a fixed cost of £50 000 for overheads, and a variable cost for each person employed. What is the variable cost? What is the total cost if the company expands to employ 45 people?

## Solution

Suppose we let:

$T$ = total cost

$O$ = overheads

$V$ = variable cost per employee

$N$ = number of people employed

We can describe the total cost as:

total cost = overheads + (variable cost $\times$ number employed)

or

$$T = O + NV$$

We know current values for $T$, $O$ and $N$, so we can find the variable cost, $V$, by rearranging the equation. If we subtract $O$ from both sides we get:

$$T - O = NV$$

Now if we divide both sides by $N$ and turn the equation around, we get:

$$V = \frac{T - O}{N}$$

Substituting the values we know for $T$, $O$ and $N$ gives:

$$V = \frac{250\,000 - 50\,000}{10} = £20\,000 \text{ a year for each employee}$$

If the company expands, we can find the new total cost $T$ by substituting the known values for $V$, $O$ and $N$ into the original equation.

$$T = O + NV = 50\,000 + (45 \times 20\,000) = £950\,000 \text{ a year}$$

## we Worked example 2.8

A manufacturer sends 1200 parts to a customer in two batches. Each unit of the first batch costs £35, while each unit of the second batch costs £37. If the total cost was £43 600 how many units were in each batch?

## Solution

Let $F$ be the number of units in the first batch, and $(1200 - F)$ be the number in the second batch. Then the total cost is:

$$35F + 37(1200 - F) = 43\,600$$

So:

$$35F + 44\,400 - 37F = 43\,600$$
$$35F - 37F = 43\,600 - 44\,400$$
$$-2F = -800$$
$$F = 400$$

So the first batch had 400 units while the second batch had $1200 - 400 = 800$ units.

## In summary

**Equations show the relationships between variables and constants. We can solve these equations to find previously unknown values. To solve an equation, we rearrange it until the unknown variable is on one side and all the known values are on the other side.**

### 2.2.3  Solving simultaneous equations

In section 2.2.2 we saw how to solve an equation and find one previously unknown value. If we want to find more than one unknown value, we have to use more than one equation. To be specific, if we want to find $n$ values we need $n$ independent equations relating them.

Suppose we know that $x + y = 3$. It is impossible to find values for both $x$ and $y$ from this one equation. But if we also know that $y - x = 1$ we have two independent equations and can find values for both $x$ and $y$ (in fact $x = 1$ and $y = 2$).

Here **independent** means that the two equations are not simply different ways of saying the same thing. For example:

$$x - y = 10$$

and

$$2x - 20 = 2y$$

are *not* independent as they are just different ways of saying the same thing.

Sets of independent equations are called **simultaneous equations.**

Suppose we have two simultaneous equations with two unknown variables. The way to solve them is to multiply one equation by a number that allows the two equations to be added or subtracted to eliminate one of the variables. After eliminating one variable, we are left with a single equation with one variable. We can solve this and then substitute into an original equation to find the value of the second variable. This sounds rather complicated, but you can follow the procedure in Worked Example 2.9.

## we Worked example 2.9

Two variables $x$ and $y$, are related by the following simultaneous equations. What are their values?

$$3y = 4x + 2 \tag{1}$$

$$y = -x + 10 \tag{2}$$

## Solution

If we multiply equation (2) by 3, we get the revised equations:

$$3y = 4x + 2 \quad \text{(as before)} \tag{1}$$

$$3y = -3x + 30 \tag{2}$$

Subtracting equation (2) from equation (1) gives:

$$3y - 3y = (4x + 2) - (-3x + 30)$$

or

$$0 = 7x - 28$$

$$x = 4$$

Now we have a value for one variable. We can substitute this into one of the original equations, say (1), to find the value of the other variable:

$$3y = 4x + 2$$

so

$$3y = 4 \times 4 + 2 = 18$$

$$y = 6$$

Finally we can make sure these values are right by checking in equation (2):

$$3y = -3x + 30$$

or

$$18 = -12 + 30$$

which is correct and confirms the solution.

We can use this method of elimination with any number of variables. If, for example, we have three unknown variables, we can manipulate these until we get two equations with two unknowns, and then do some more manipulation to get one equation with one unknown. Then we can solve this equation and substitute values into earlier equations to give the final solution.

## we Worked example 2.10

Solve the simultaneous equations:

$$2x + y + 2z = 10 \tag{1}$$

$$x - 2y + 3z = 2 \tag{2}$$

$$-x + y + z = 0 \tag{3}$$

## Solution

We can start by using equations (2) and (3) to eliminate the variable $x$ from equation (1).

- Multiplying equation (2) by 2 gives:      $2x - 4y + 6z = 4$

  Subtracting this from equation (1) gives:      $5y - 4z = 6$       (4)

- Multiplying equation (3) by 2 gives:      $-2x + 2y + 2z = 0$

  Adding this to equation (1) gives:      $3y + 4z = 10$       (5)

Now we have two equations, (4) and (5), with two unknowns, $y$ and $z$.

Adding equations (4) and (5) together gives:    $8y = 16$, so $y = 2$

We can substitute this value for $y$ in earlier equations to find values for $x$ and $z$. If we substitute $y = 2$ into equation (4) we get:

$$10 - 4z = 6 \quad \text{or} \quad z = 1$$

Now we can substitute these values for $y$ and $z$ into equation (1) to get:

$$2x + 2 + 2 = 10 \quad \text{or} \quad x = 3$$

This gives the complete solution with:

$$x = 3 \quad y = 2 \quad z = 1$$

We can check these values by substituting in the original equations.

$2x + y + 2z = 10 \quad \text{or} \quad 6 + 2 + 2 = 10$       (1)

$x - 2y + 3z = 2 \quad \text{or} \quad 3 - 4 + 3 = 2$       (2)

$-x + y + z = 0 \quad \text{or} \quad -3 + 2 + 1 = 0$       (3)

As you can see from Worked Example 2.10, the arithmetic for this method of eliminating variables can become rather tedious. But we know that computers can do the calculations. Figure 2.1 shows an example of a printout from a program solving four equations.

## In summary

**To find values for $n$ unknown variables we need $n$ independent simultaneous equations. We can solve these equations using a process of elimination and substitution. Computers will do these calculations automatically.**

---

PROGRAM SOLVING: LINEAR SIMULTANEOUS EQUATIONS

SUMMARY OF OUTPUT DATA

Number of variables:    4
Names of variables:    A    B    C    D

Equations:
1. 1A + 1B + 1C + 1D = 10
2. 2A + 0B − 4C − 1D = 2
3. 0A + 1B + 2C − 1D = 3
4. 0A + 1B + 0C + 1D = 5

SOLUTION FOUND

SUMMARY OF OUTPUT DATA

Variable values:
    A = 4
    B = 3
    C = 1
    D = 2

Check of calculations:
1. $1 \times 4 + 1 \times 3 + 1 \times 1 + 1 \times 2 = 10$
2. $2 \times 4 + 0 \times 3 − 4 \times 1 − 1 \times 2 = 2$
3. $0 \times 4 + 1 \times 3 + 2 \times 1 − 1 \times 2 = 3$
4. $0 \times 4 + 1 \times 3 + 0 \times 1 + 1 \times 2 = 5$

END OF PROGRAM

---

**Figure 2.1** Printout from computer program solving simultaneous equations.

## 2.2.4 Powers and roots

When we multiply a number, $b$, by itself, the result is '$b$ squared' or '$b$ to the power 2', which is written as $b^2$. When 3 is multiplied by itself the result is '3 squared' which is written as $3^2$, and this equals $3 \times 3$ or 9. Similarly:

$b$ squared $= b \times b = b^2$

$b$ cubed $= b \times b \times b = b^3$

$b$ to the fourth $= b \times b \times b \times b = b^4$

and, in general,

$b$ to the power $n = b \times b \times b \times \ldots$ ($n$ times) $= b^n$

$b$ raised to the power 1 is just $b$; one slightly unusual result is that $b$ raised to the power zero equals 1.

Then if $b = 2$ we have:

$$2^0 = 1 \quad 2^1 = 2 \quad 2^2 = 4 \quad 2^3 = 8 \quad 2^4 = 16 \quad \text{etc.}$$

If $b = 3$ we have:

$$3^0 = 1 \quad 3^1 = 3 \quad 3^2 = 9 \quad 3^3 = 27 \quad 3^4 = 81 \quad \text{etc.}$$

We can raise variables to negative powers, using the general rule:

$$b^{-n} = \frac{1}{b^n}$$

So,

$$2^{-2} = \frac{1}{2^2} = \tfrac{1}{4} = 0.25$$

and

$$3^{-3} = \frac{1}{3^3} = \tfrac{1}{27} = 0.037$$

We can also raise variables to fractional powers. We rarely do this kind of calculation by hand, but we can still give a couple of definitions. First:

$$b^{0.5} = b^{1/2} = \sqrt{b}$$

Remember that $\sqrt{\phantom{x}}$ is the sign for a square root, so $9^{0.5} = 9^{1/2} = \sqrt{9} = 3$. In the same way:

$b^{0.33} = b^{1/3} =$ the cube root of $b$

$b^{0.25} = b^{1/4} =$ the fourth root of $b$

$b^{0.2} = b^{1/5} =$ the fifth root of $b$, and so on

## we Worked example 2.11

What are the values of (a) $4^{-3}$, (b) $25^{1/2}$, (c) $17.1^0$?

## Solution

Using the rules defined above:

(a)  $4^{-3} = 1/(4^3) = 1/64 = 0.0156$

(b)  $25^{1/2} = \sqrt{25} = 5$

(c)  $17.1^0 = 1$   (as anything raised to the power 0 equals 1)

We can use powers to get a convenient notation for very large or very small numbers. **Scientific notation** is used to represent any number in the format:

$$a \times 10^b$$

where:

$a$ is a number between 0 and 10

$b$ is the appropriate power of 10

This notation uses the fact that $10^1 = 10$, $10^2 = 100$, $10^3 = 1000$ and so on. Then we can write:

12 as $1.2 \times 10^1$   (i.e. $1.2 \times 10$)

1200 as $1.2 \times 10^3$   (i.e. $1.2 \times 1000$)

120 000 as $1.2 \times 10^5$   (i.e. $1.2 \times 100\,000$)

1 380 197.892 as approximately $1.38 \times 10^6$

the value of the UK's annual exports as about £$2 \times 10^{11}$

The superscript of the 10 shows the number of places that the decimal point is moved. Similarly, $10^{-1} = 0.1$, $10^{-3} = 0.001$ and $10^{-6} = 0.000001$, so we can write:

0.12 as $1.2 \times 10^{-1}$   (i.e. $1.2 \times 0.1$)

0.0012 as $1.2 \times 10^{-3}$   (i.e. $1.2 \times 0.001$)

0.000 012 as $1.2 \times 10^{-5}$

0.000 004 29 as $4.29 \times 10^{-6}$

We will not use this notation very much, but it is sometimes useful for describing large or small numbers.

## In summary

**Superscripts show that values are raised to powers. Negative superscripts show division, and fractional superscripts show roots. These can all be used in a scientific notation.**

## Chapter review

This chapter looked at some basic arithmetic and algebra, which we will use in the rest of the book. In particular, the chapter described:

- calculations using integers, decimals, fractions, and percentages;
- rounding numbers;
- the use of algebra;
- how to solve an equation with one unknown;
- how to solve simultaneous equations;
- superscripts to describe powers;
- scientific notation.

# Problems

2.1 Evaluate the following:
   (a) $-12 \times 8$
   (b) $-35/(-7)$
   (c) $(24 - 8) \times (4 + 5)$

   (d) $((18 - 4)/(3 + 9 - 5))$

2.2 Simplify the fractions:
   (a) $\frac{3}{5} + \frac{1}{2}$
   (b) $\frac{3}{4} \times \frac{1}{6}$
   (c) $\frac{3}{4} - \frac{1}{8}$
   (d) $-\frac{18}{5} \div \frac{6}{25}$

2.3 What are the answers to Problem 2.2 as decimal fractions?

2.4 (a) What is 23/40 as a percentage?
   (b) What is 65% as a fraction?
   (c) What is 17% as a decimal?

2.5 What is 1037/14 to (a) three decimal places, (b) one decimal place, (c) two significant figures, (d) one significant figure?

2.6 In one exam 64 people passed and 23 failed; in a second exam 163 people passed and 43 failed. Which exam was the more difficult to pass?

2.7 A shopkeeper buys an item from a wholesaler and sells it to customers. What equation describes the shopkeeper's profit?

2.8 A club has £1515 to spend on footballs. Match balls cost £35 each while practice balls cost £22 each. The club buys 60 balls each year. How many of each type does it buy to exactly match the budget?

2.9 Salmak Leisure finds that 30% of its costs are direct labour. Each week raw materials cost £1000 more than twice this amount, and there is an overhead of 20% of direct labour costs. What are the weekly costs to the company?

2.10 Solve the following simultaneous equations:
   (a) $a + b = 3$  and  $a - b = 5$
   (b) $2x + 3y = 27$  and  $3x + 2y = 23$

2.11 A company finds that one of its productivity measures is related to the number of employees, $e$, and the production, $n$, by the following equations:

   $10n + 3e = 45$
   $2n + 5e = 31$

   What are the current values for $e$ and $n$?

2.12 What are the values of (a) $3^4$, (b) $9^{0.5}$, (c) $4^{2.5}$?

part

# 2

# *Data collection and description*

In Part One we looked at some fundamental ideas of quantitative methods. Now, Part Two shows how to collect and describe data.

There are three chapters in Part Two. Chapter 3 describes how to collect data, then Chapters 4 and 5 show how to summarise data and present them in different ways.

# 3 Collecting data

## Contents

## Chapter outline

This chapter describes different types of data, and shows how to collect them. After reading the chapter you should be able to:

- understand the importance of data collection
- consider the amount of data to collect
- classify data in several ways
- understand the use of sampling
- choose different types of sample
- collect data in different ways
- discuss the design of questionnaires

# 3.1 Introduction

## 3.1.1 Why collect data?

There is a difference between data and information. **Data** are the raw numbers or facts that we must process to give useful **information**. For example, 78, 64, 36, 70 and 52 are data; processing these gives the information that five student reports have an average mark of 60.

You can see why data collection is important from the following argument.

---

**Importance of data collection**

- Managers make decisions within an organisation.
- They can only make good decisions if they have the right information.
- To get this information they have to collect and analyse data.

---

All decisions within an organisation are based on data collection. Figure 3.1 shows the process of moving from data collection to decisions. In the rest of this chapter we are going to see how to collect reliable data.

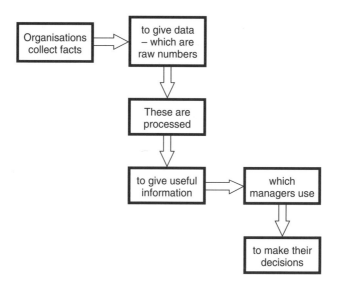

**Figure 3.1** The importance of data collection.

## In summary

**Managers need reliable information to make decisions. To get this information, they must collect and analyse data. This means that data collection is an important function in every organisation.**

## 3.1.2 How much data should you collect?

You should always collect data for a specific purpose. This means that you look at a problem, and then decide what data to collect. This probably seems obvious, but many people do it the other way round: they collect some data and then see what they can do with it. As you can imagine, this gives all sorts of problems with data that are missing, or in the wrong form, out of date, or are somehow unreliable.

Even when you know what data you want, you have to decide how much to collect. There is often a huge amount of data that you could collect, and if you look through the World Wide Web you can find thousands of pages of data on almost anything. But collecting and processing data costs money, so you should always limit yourself to data that are relevant and useful.

In theory, there is an optimal amount of data to collect for any problem. If you imagine the marginal cost of data as the cost of collecting the last unit, then the marginal cost increases with the amount of data collected. Finding some general data about, say, Canadian Pacific Railway is easy (it runs trains, employs staff, and so on). To get more detailed data we would have to go to a specialised library (to find the different types of trains it runs, or the staff of different grades it employs). To get even more detailed data we would have to search Canadian Pacific's own records (to find the wage bill for each grade of employee). To get even more detailed data would need a special survey (to ask how each grade of employee feels about their work conditions). So, collecting more data becomes more difficult and expensive.

On the other hand, as you collect more data its marginal benefit – which is the benefit of the last unit of data collected – will fall. It is very useful to know that Canadian Pacific Railway runs a rail service, but it is much less useful to know how different grades of employee feel about their working conditions. We can show this general pattern in Figure 3.2.

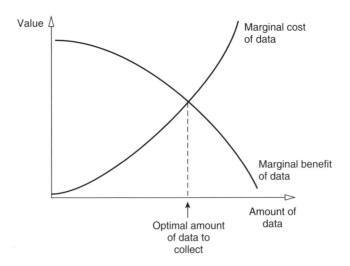

**Figure 3.2**   Finding the optimal amount of data to collect.

The optimal amount of data comes from the point where the marginal cost equals the marginal benefit. If you collect more data than this, its benefit is less than its cost and you are wasting money. If you collect less data, its benefit is more than its cost and you are missing potential benefits.

In practice, it is very difficult to measure these marginal costs and benefits, so most people use their experience and judgement to collect the amount of data that seems reasonable.

### In summary

**Data collection and analysis is expensive. We should only collect data when we know how they will be used. In principle, there is an optimal amount of data for any purpose.**

## 3.2 Types of data

There are many types of data, and each can be collected in a different way. We have already said that data can be either qualitative or quantitative. It is much easier to collect and analyse quantitative data, so you should use this whenever possible. We can extend this description by saying how well data can be measured.

- **Nominal data** cannot really be measured. The fact that a company is a manufacturer, or a country has a centrally planned economy, or a cake has cream in it, are examples of nominal data. The most common analysis of nominal data defines a number of different categories and says how many observations fall into each. A survey of companies in a particular area might show that there are 7 manufacturers, 16 service companies and 5 in primary industries. The order in which we list the categories does not have any significance.

- **Ordinal data** are one step more quantitative, as we can rank the different categories of observations. Sweaters, for example, are extra large, large, medium, small or extra small. Describing a sweater as medium tells us something, but really all it says is that a medium sweater is smaller than a large one, but larger than a small one . You will often meet this in consumer surveys which ask, 'How strongly do you agree with this statement, using a scale of 1 to 5?' Now data can be put into different categories, and the order of these categories is important.

- **Cardinal data** have some attribute that can be directly measured. We can weigh a bag of chocolates, measure the time to do a job, find the temperature in an office. Cardinal data are the easiest to analyse and are the most relevant to quantitative methods. There are two types of cardinal data:

  - **Discrete data** can only take integer values. If you ask people how many children they have, the answer is 0, 1, 2 or some other integer number. Similarly, the number of cars owned, machines operated, shops opened and people employed, are discrete.

- **Continuous data** can take any value and are not restricted to integers. For example, the weight of a bag of biscuits that might be 256.312 grammes. Similarly, the time taken to do a job, the length of metal bars, the area covered in carpet and the height of flagpoles are continuous.

Another way of classifying data depends on how they are collected. If you want some data for a particular purpose, you can use either primary or secondary data.

- **Primary data** are collected by the organisation itself for the particular purpose.

- **Secondary data** are collected by other organisations or for other purposes.

Any data that are not collected by the organisation for the specified purpose are secondary data. These may be published by other organisations, available from research studies, published by the government, and so on. They might also be collected by the organisation itself but for another purpose.

Primary data have the benefits of exactly fitting the need and of being up to date and reliable. Secondary data though, are much cheaper, easier and faster to collect. They can also come from sources that are not generally accessible. For example, companies will often answer a survey from the government, Confederation of British Industry, or group of students, but they will not answer questions from another company.

## In summary

**Data of different types are collected in different ways. We can classify data in several ways, including quantitative/qualitative, nominal/ordinal/cardinal, discrete/continuous and primary/secondary.**

# 3.3 How to choose a sample

## 3.3.1 Purpose of sampling

Most data are collected using a **sample**. This means that you collect data from a representative group of people or things, and use this sample to estimate the characteristics of all people or things.

Suppose a company is about to launch a new product and wants to find the likely sales. It could:

- either ask every person in the country how much of the product they will buy – this is a **census**;

- or ask some people how much of the product they will buy, and use their answers to estimate the demand from the whole population – this is a **sample**.

A census gives accurate results, but it takes a lot of time and is expensive. Using a sample is quicker and cheaper, but it may be less accurate. The aim of sampling is to get reliable results using only a small part of the whole population. Notice that we are using population in its statistical sense of a set of items that share some common characteristics.

> When collecting data:
>
> - the **population** is all people or things that could give data
> - the **sample** is the people or things that actually do give data

If the Post Office wants to see how long it takes to deliver first class letters, the population is all letters that are posted first class. The population for users of an electronic game might be all girls between the ages of 10 and 14. When testing a product's quality, the population is all units of the product that are made.

When you collect data, it is important to use the right population. This may seem obvious, but it can be quite difficult. The population for the electronic game mentioned above might actually include boys aged 10 to 14 and some of their parents. For a survey of student opinion the population is clearly all students – but does this mean full-time students only, or does it include part-time, day-release and distance-learning students; what about students doing block-release courses during their period of work, school students and those studying but not enrolled in courses? If you make a mistake defining the population, all the work you do afterwards is wasted.

Even when you identify the right population, it may be difficult to actually collect data. Suppose you identify a population as all people who used a particular supermarket during the past two years: how could you get a list of these people? Sometimes, if you are lucky, it is fairly easy to list the population. If the population is houses with telephones, you can find most of these from telephone directories. Such lists of the population are called **sampling frames**, and you can find them in electoral registers, club membership lists, credit rating agencies and so on.

When you have found the population – and can actually identify it – you have to choose:

- a sample size – which must be big enough to represent the whole population, but small enough to be practical;

- a method of choosing this sample.

We return to the problem of finding a sample size in Chapter 12, but the next section describes some ways of choosing the sample.

## In summary

**Data collection often uses sampling. This takes data from a representative sample, and uses the results to estimate data for the population. Sampling is used when a census is too expensive, time consuming or impractical.**

## 3.3.2 Types of sample

The most common ways of choosing a sample are set out below.

- Census
- Random sample
- Systematic sample
- Quota sample
- Stratified sample
- Multi-stage sample

**Census**

If the population is small and the results are important, it might be worth using a census to collect data from every member of the population. The UK government carries out a population census of this kind every ten years. A census has the benefit of giving very accurate data, but is too expensive and time consuming for most purposes.

**Random sample**

A sample should give a true picture of the population. The easiest way of arranging this is with a **random sample**. This means that every member of the population has exactly the same chance of being chosen for data collection – but it does not mean that the sample is disorganised or haphazard. If you want some data about tinned soup, you could go to a supermarket and buy the first dozen tins of soup you see. This is haphazard, but it certainly is not random – as tins in other supermarkets have no chance of being chosen.

There are several ways of choosing random samples from a population. A club with a few members can pick names from a hat. The National Lottery uses machines to choose six random, numbered balls. The most common way of choosing random samples is to get a computer to generate sets of random numbers. These are simply strings of random digits, like 4, 9, 3, 1, 1, 7, 4, etc. Then if, for example, you want to collect data from a random sample of people visiting an office, you could stand by the door and interview the fourth person to pass, then the ninth person after that, then the third after that, then the first after that, and so on.

Some samples that seem to be random are not. If you decide to save time and simply write down a series of numbers, they are not random – as you may have a preference for even numbers or for sequences that are easy to type on a keyboard. Similarly, interviewers in the street cannot choose people at random: they are more likely to approach people that they find attractive and more likely to avoid people they find unattractive, intimidating, people in an obvious hurry or people in groups.

# we Worked example 3.1

In the last financial year Alamos Pharmaceuticals was sent 10 000 invoices. The auditors do not have time to look at each invoice, so they want a random sample of 200. How can they choose the sample?

## Solution

The auditors can start by forming the sampling frame. For this they list the invoices and number them 0000 to 9999. Then they can use a computer to generate a set of 200 random numbers, each with four digits. One set is:

4271   6845   2246   9715   4415   0330   8837   etc.

Choosing these invoices gives a random sample.

A random sample will, in the long run, accurately represent the whole population. If a sample does not represent the population exactly, it is **biased**. Unfortunately, it takes a large random sample to make sure that there is no bias. If the sample is too small, it might include a few unusual results that give bias.

Suppose that you want to find the average height of people living in a city. If you take a random sample of only three people, you might get unusual results and find the average height is 6 ft 10 in. To get a reliable answer you need a sample of several hundred people. But there is a way of using a smaller sample and still getting results that are not biased. This uses a structured or non-random sample.

**Systematic sample**

The easiest way of organising a non-random sample is to collect data at regular intervals. We could, for example, weigh every tenth unit from a production line, or interview every twentieth person using a service.

With a random sample every member of the population has the same chance of being chosen. A systematic sample though, is not random. If we choose every tenth member, then members 11, 12, 13, etc. have no chance of being chosen. Unfortunately, there are times when this regularity introduces bias. Checking the contents of every twentieth bottle filled in a bottling plant may be misleading if every twentieth bottle is filled by the same head on the filling machine. Collecting data from every thirtieth person leaving a bus station may give bias if buses hold an average of 30 people, and we are always interviewing the last people to get off.

## we Worked example 3.2

A production line makes 5000 units a day. How can quality controllers check a systematic sample of 2% of production?

## Solution

To sample 2% of production needs $5000 \times 0.02 = 100$ checks a day. So a systematic sample would check every $5000/100 = 50$th unit.

**Quota sample**

Another way of structuring samples is to make sure that the overall sample has the same characteristics as the population. Suppose you want to see how people will vote in an election. You can look at the population figures – where the population is those people who can vote – and see what proportions have various characteristics. Then you can choose a sample with exactly these characteristics. If the population has 53% women, 32% of whom are over 50 years old, then the sample will also have 53% women, 32% of whom are over 50 years old.

To get these proportions, each interviewer is given a quota of people with different characteristics to interview – perhaps 12 women who are single, between 20 and 30 years old, have full-time professional jobs, and so on. The interviewer chooses people at random with these characteristics. Although the actual choice of people is random, the process is not random, as an interviewer who has already filled the quota of one category of people does not interview anyone else in this category, so those people have no chance of being chosen.

## we Worked example 3.3

There are 56 300 people who can vote in an election. Census records show that the population has the characteristics shown in Table 3.1.

An opinion poll wants to interview 1200 people to see how they will vote. How many people should be in each category?

**Table 3.1**

| | | |
|---|---|---|
| Age | 18 to 25 | 16% |
| | 26 to 35 | 27% |
| | 36 to 45 | 22% |
| | 46 to 55 | 18% |
| | 56 to 65 | 12% |
| | 66 and over | 5% |
| Sex | Female | 53% |
| | Male | 47% |
| Social class | A | 13% |
| | B | 27% |
| | C1 | 22% |
| | C2 | 15% |
| | D | 23% |

## Solution

The sample should contain exactly the same proportion in each category as the population. So, 16% (or 192 people) should be aged 18 to 25. Of these 192 people, 53% (or 102) should be women. Of these 102 women, 13% (or 13) should be in social class A. Similarly, 5% (or 60 people) should be over 66 years old; 47% (or 28) of these should be male, and 23% of these (or 6 people) should be in social class D. Using the same reasoning for all other combinations gives the quotas shown in the spreadsheet in Figure 3.3.

| | A | B | C | D | E | F | G | H |
|---|---|---|---|---|---|---|---|---|
| 1 | **Quota samples** | | | | | | | |
| 2 | | | | | | | | |
| 3 | | | **Age** | | | | | |
| 4 | | | **18 to 25** | **26 to 35** | **36 to 45** | **46 to 55** | **56 to 65** | **66 and over** |
| 5 | Female | **A** | 13 | 22 | 18 | 15 | 10 | 4 |
| 6 | | **B** | 27 | 46 | 38 | 31 | 21 | 9 |
| 7 | | **C1** | 22 | 38 | 31 | 25 | 17 | 7 |
| 8 | | **C2** | 15 | 26 | 21 | 17 | 11 | 5 |
| 9 | | **D** | 23 | 39 | 32 | 26 | 18 | 7 |
| 10 | | | | | | | | |
| 11 | Male | **A** | 12 | 20 | 16 | 13 | 9 | 4 |
| 12 | | **B** | 24 | 41 | 34 | 27 | 18 | 8 |
| 13 | | **C1** | 20 | 34 | 27 | 22 | 15 | 6 |
| 14 | | **C2** | 14 | 23 | 19 | 15 | 10 | 4 |
| 15 | | **D** | 21 | 35 | 29 | 23 | 16 | 6 |

**Figure 3.3** Spreadsheet for quotas in Worked Example 3.3.

### Stratified sample

When there are distinct groups or strata in a population, it can be useful to have representatives from each stratum in the sample. Suppose you want data from different types of companies, including manufacturers, transport operators, retailers, wholesalers and so on. A particular area may only have one or two manufacturers, but it is still important to get their views. A stratified sample divides the population into strata, and chooses a random sample from each. This means that small strata are over-represented, and any conclusions have to bear this in mind.

### Multi-stage sample

Suppose you want a sample of people who subscribe to a certain magazine. If you take a random sample, you will probably find that the subscribers live all over the country and you would need to do a lot of travelling to meet them. A cheaper option uses **multi-stage sampling**. This divides the country into geographical regions – such as independent television regions. Then you choose some of these regions at random and divide them into smaller areas, perhaps parliamentary constituencies or local authority areas. Then you choose some of these areas at random and divide them into even smaller areas, perhaps towns or parliamentary wards. You continue doing this until you have small enough areas to do a survey without much travelling.

## In summary

**You can get samples in many different ways, including census, random, systematic, quota, stratified and multi-stage sampling.**

# 3.4 Ways of collecting data

## 3.4.1 Types of survey

When we have chosen a suitable sample – and for simplicity we will assume that we are conducting a survey of people – the next stage is to actually collect data. The Gallup organisation says there are five reasons for doing surveys:

- To see if people are aware of an issue ('Do you know of any plans to develop...?')

- To find general feelings about an issue ('Do you think this development is beneficial...?')

- To get views about specific points in an issue ('Do you think this development will affect...?')

- To get reasons for people's views ('Are you against this development because...?')

- To see how strongly people hold these views ('On a scale of 1 to 5, how strong are your feelings about this development...?')

To collect these views we can use any of the following.

- Observation
- Personal interviews
- Telephone interviews
- Postal surveys
- Panel surveys
- Longitudinal surveys

### Observation

If you want data from machines, animals, files or other objects, the only way of collecting them is by direct observation. Even when the population is people, you can get more reliable results from observation, as people tend to give the answer that they think they ought to give or that the interviewer wants, rather than the true answer.

### Personal interview

Personal interviews are the most reliable way of collecting data from people. They also get a high response rate, with only about 10 per cent of people refusing to answer questions.

Collecting data by personal interview seems easy, but interviewers must be trained so that they do not explain the questions, or help people with answers, or encourage certain answers by their expression, tone of voice, or comments. These would all introduce the interviewer's bias to the answer.

### Telephone interview

About 90% of houses have a telephone, so this is a popular way of asking questions. It has the advantages of being cheap, easy to organise, requires no travel by interviewers and gets a high response rate. On the other hand, it has the disadvantage of introducing bias (as it only uses people with telephones), allowing no personal observation of respondents, and annoying people who object to the intrusion of their homes.

### Postal survey

Sending a questionnaire through the post has the advantage of being cheap and easy to organise. You can use a very large sample – but will often get few replies. Even a good postal survey gets replies from less than 20% of people questioned. You can increase this response rate by making sure the questionnaire is short and easy to complete, sending it to a named individual, adding a covering letter to describe the purpose of the survey, enclosing a prepaid return envelope, promising to keep the replies anonymous, sending a follow-up letter for slow replies, promising a summary of results, and offering some reward for completion.

Unfortunately, postal surveys often get biased results. Asking people for their views on, say, a holiday will get a higher response rate from those who had a bad experience and want to complain, than those who had a good experience.

### Panel survey

Panel surveys usually monitor changes over time. You start with a panel of respondents, and periodically ask them a series of questions. Then you can monitor the political views of a panel during the life of a government, or see how the panel becomes aware of a product during an advertising campaign. Panel surveys are expensive and difficult to administer, so they use fairly small samples.

### Longitudinal survey

This is an extension of a panel survey which monitors a group of respondents over a long period. One television company has, for example, been monitoring the progress of a group of children for the past 35 years. Maintaining this kind of extended survey needs a lot of resources, so longitudinal surveys are generally limited to studies of health, physical and sociological changes.

## we Worked example 3.4

In the late 1970s Atlantic Cars was about to sell a new high performance sports car. It made a couple of prototypes and asked a market research company to obtain reactions. The reports that came back were very positive, so Atlantic spent several million pounds on new machinery and started making the car. Unfortunately, sales were lower than expected. Atlantic was soon losing money and was taken over by another company. What might they have done wrong?

## Solution

Atlantic's problems started with an over-optimistic view of sales from the market research company. This company had recruited twenty part-time interviewers from a local college. Unfortunately, all the interviewers were female, and they chose to collect data from young, male students. Anyone showing particular interest in the car was offered a short test drive. Not surprisingly, the interviewers reported a lot of interest. In practice, of course, this was never translated into sales.

You may think that this is an extreme case, but many surveys are done badly. In 1992, for example, every survey of voting intentions showed that the Labour Party would win the general election. On the day, they lost by a considerable margin.

## In summary

**After identifying a sample, we have to collect data. There are several ways of doing this, including observation, personal interview, telephone interview, postal survey, panel survey and longitudinal survey.**

## 3.4.2 Design of questionnaires

Data collection often uses questionnaires – which are surprisingly difficult to design. Even small changes – like using different coloured paper, or different fonts for the questions, or describing 'not winning' instead of 'losing' – can give completely different answers. Some guidelines can be given for questionnaires, but most of these are common sense.

- Ask a series of short, simple questions phrased in everyday terms. Put these in a logical order.

- Make questions simple and easy to understand: if people do not understand the question, they will give any convenient answer rather than the true one.

- Make questions clear, unambiguous and without conditional clauses.

- Be very careful with the phrasing of questions. Even simple changes in wording give different results – so a medical treatment with a 60% success rate is better than a treatment with a 40% failure rate.

- Avoid leading questions like, 'Do you agree with the common view that BBC television programmes are better than commercial ones?'. This encourages people to conform rather than give true answers.

- Use phrases that are as neutral as possible. 'Do you like this cake?' can be rephrased as, 'Say how you feel about the taste of this cake on a scale of 1 to 5'.

- Remember that respondents are not always objective. The question, 'Do you think prison sentences should be used to deter speeding drivers?' will get a different answer from, 'If you were caught speeding do you think you should go to prison?'

- Phrase all personal questions carefully. 'Have you retired from paid employment?' might get better replies than the more sensitive, 'How old are you?'.

- Do not start questions by warning clauses. Hardly anyone will answer a question that starts, 'We hope you do not mind answering this question, but will understand if you do not want to…'.

- Avoid vague questions like, 'Do you usually buy more meat than vegetables?'. This raises questions about what is usual, what is more, and whether frozen meals count as meat or vegetables.

- Ask positive questions like, 'Did you buy a Sunday newspaper last week?', rather than the less definite, 'How has the number of Sunday newspapers you read changed in the past few years?'

- Avoid hypothetical questions like, 'How much would you spend on life insurance if you won £5 million on the National Lottery?'

- Avoid asking two or more questions in one, like, 'Do you think this development should go ahead because it will increase employment in the area and improve facilities?'. This will get confused answers from people who think the development should not go ahead, or those who think it will increase employment but not improve facilities.

- Make the questionnaire as short as possible and do not ask irrelevant questions. It is often tempting to add an extra question or two, but people will not bother answering a poor or long questionnaire.

- Open questions, such as 'Have you any other comments to make?', allow general comments, but they favour the articulate and quick-thinking.

- Ask questions with pre-coded answers, so that respondents choose the most appropriate answer.

- Be prepared for unexpected effects, such as sensitivity to the colour and format of the questionnaire, or different interviewers getting different replies.

- Always run a pilot survey before starting the whole survey. This will show any problems and allow you to improve the questionnaire.

## In summary

**Designing a good questionnaire is difficult and needs a lot of thought. There are several guidelines for questionnaires. You should always run a pilot survey to sort out any problems.**

## 3.4.3 Non-responses

About 80% of questionnaires sent by post and 10% of personal interviews will get no response. Reasons why people do not respond include:

- they cannot answer the questions – perhaps because of language difficulties, illness, or they simply do not know the answer;

- they were out when the interviewer called – you can reduce this problem by careful timing of calls and by making second visits;

- they were away for some longer period – holiday and business trips make surveys in summer more difficult;

- they have moved house – it is rarely worth following up a new address;

- they refuse to answer – only a few people refuse to answer on principle, but you cannot do anything about these.

You might be tempted to ignore non-responses and assume that the data collected are typical of the sample. In other words, you assume that the respondents properly represent the sample, which in turn properly represents the population. This may, however, not be true. In an extreme case, a postal questionnaire might try to see how fluently people can read – in the same way that people who have reading difficulties can pick up packages of information when they visit their local library. You can also get biased replies by asking questions like, 'Do you have a computer?' People who answer 'No' will probably not be interested enough to complete the rest of the questionnaire, so the responses are biased towards people who have computers.

To avoid this kind of bias, you should follow up non-respondents with another visit, telephone call or letter. Then you have to examine non-respondents to make sure they do not share some common characteristic that is missing in respondents.

## In summary

**Most surveys can expect some non-respondents. You should examine these carefully to make sure that they do not introduce bias to the data.**

# Chapter review

This chapter discussed some aspects of data collection. In particular it:

- showed that managers need information, and that this relies on data collection;
- mentioned the best amount of data to collect;
- classified data according to qualitative/quantitative, nominal/ordinal/cardinal, discrete/continuous and primary/secondary;
- showed how data collection uses samples from populations;
- discussed sampling methods, including census, random, systematic, quota, stratified and multi-stage samples;
- listed ways of collecting data, including observation, personal interview, telephone interview, postal survey, panel survey and longitudinal survey;
- gave some guidelines for questionnaire design;
- mentioned non-responses.

Now we can collect together these ideas and summarise the stages in data collection as follows.

## Stages in data collection

- See why you need the data.
- See what kind of data you want.
- Look for secondary data.
- Define the population to give primary data.
- Find the best sampling method and sample size.
- Identify an appropriate sample.

- Design a questionnaire or other method of collection.
- Train any interviewers, observers or experimenters.
- Run a pilot study.
- Do the main study.
- Do any necessary follow-up, such as contacting non-respondents.
- Analyse and present the results.

## Problems

3.1 Collect some government statistics to show how the gross national product (GNP) has changed over the past twenty years.

3.2 What population could give data on the following?
(a) Likely sales of a computer game.
(b) Problems facing small shopkeepers.
(c) Proposals to close a shopping area to all vehicles.
(d) Changes in house prices.
(e) European car production.

3.3 An auditor wants a sample of 300 invoices from 9000 available. How could she get this?

3.4 The characteristics of the readership of a Sunday newspaper are listed in Table 3.2. What would be the quotas for a sample of size 2000?

**Table 3.2**

| | | |
|---|---|---|
| Age | 16 to 25 | 12% |
| | 26 to 35 | 22% |
| | 36 to 45 | 24% |
| | 46 to 55 | 18% |
| | 56 to 65 | 12% |
| | 66 to 75 | 8% |
| | 76 and over | 4% |
| Sex | Female | 38% |
| | Male | 62% |
| Social class | A | 24% |
| | B | 36% |
| | C1 | 24% |
| | C2 | 12% |
| | D | 4% |

3.5     Run a survey to find the opinions of your colleagues on proposals to restrict smoking in public places.

3.6     Design a questionnaire to collect data on the closure of a shopping area to all vehicles.

3.7     Find a copy of a recent survey by the Consumers' Association (or any equivalent organisation). Describe the data collection used.

3.8     Do a survey of the use of information technology by companies working in your area. How would you choose a sample of companies for this?

# CS Case study – Natural Biscuits

Natural Biscuits sell a range of products to health food shops around the country. It divides the UK into thirteen geographical regions based around major cities. The populations, number of shops stocking its goods and annual sales in each region last year are shown in Table 3.3

**Table 3.3**

| Region | Population (millions) | Shops | Sales (£000) |
|---|---|---|---|
| Greater London | 8,130 | 94 | 240 |
| Birmingham | 1,205 | 18 | 51 |
| Glasgow | 870 | 8 | 24 |
| Leeds | 853 | 9 | 18 |
| Sheffield | 641 | 7 | 23 |
| Liverpool | 580 | 12 | 35 |
| Bradford | 556 | 8 | 17 |
| Manchester | 541 | 6 | 8 |
| Edinburgh | 526 | 5 | 4 |
| Bristol | 470 | 17 | 66 |
| Coventry | 372 | 8 | 32 |
| Belfast | 365 | 4 | 15 |
| Cardiff | 336 | 4 | 25 |

Natural Biscuits is about to introduce a 'Vegan Veggie Bar' which is made from a combination of nuts, seeds and dried fruit, and is guaranteed to contain no animals' products. The company wants to run a market survey to find likely sales of the bar. It already sells 300 000 similar bars each year, with an average price of 40p and a profit of 7.5p per bar. An initial survey of 120 customers in three shops earlier this year gave the characteristics of customers for these bars, shown in Table 3.4.

**Table 3.4**

| | | |
|---|---|---|
| Sex | Female | 64% |
| | Male | 36% |
| Age | Less than 20 | 16% |
| | 20 to 30 | 43% |
| | 30 to 40 | 28% |
| | 40 to 60 | 9% |
| | More than 60 | 4% |
| Social class | A | 6% |
| | B | 48% |
| | C1 | 33% |
| | C2 | 10% |
| | D | 3% |
| Vegetarian | Yes | 36% (5% vegan) |
| | No | 60% |
| | Other response | 4% |
| Reason for buying | Like the taste | 35% |
| | For fibre content | 17% |
| | Never tried before | 11% |
| | Help diet | 8% |
| | Other response | 29% |
| Regular buyer of bar | Yes | 32% |
| | No | 31% |
| | Other response | 37% |

Natural Biscuits wants as much information as possible about likely sales, but also wants to limit costs to reasonable levels. It costs about £10 to interview a customer personally, while a postal or telephone survey costs £5 a response.

How would you design Natural Biscuits' data collection? What would be the timing and costs?

# 4 Using diagrams to present data

## Contents

## Chapter outline

There are two ways of summarising data: using either diagrams or numbers. This chapter shows how to use diagrams, and the next chapter uses numbers. After reading the chapter you should be able to:

- discuss the reasons for reducing data
- use tables to present data
- work with frequency distributions
- draw graphs to show the relationship between variables
- design pie charts, bar charts and pictograms
- use histograms

# 4.1 Data reduction

The previous chapter described the difference between data and information. Data are the raw numbers or facts that we must process to give useful information. So 78, 64, 36, 70 and 52 are data. We can process this to give the information that five student reports got an average mark of 60 per cent.

Most people can deal with small amounts of numerical data. We are happy to say, this building is 60 metres tall, a car travels 40 miles on a gallon of petrol, and one political party has 6 per cent more support than another. If there is a lot of data, though, we get swamped by the details. Suppose that the weekly sales of a product over the past year are:

```
51  60  58  56  62  69  58  76  80  82  68  90  72
84  91  82  78  76  75  66  57  78  65  50  61  54
49  44  41  45  38  28  37  40  42  22  25  26  21
30  32  30  32  31  29  30  41  45  44  47  53  54
```

In this format these figures are so boring that you have probably skipped straight over them. The problem is that raw data overwhelm us with detail, when really we want an overall picture. So we need some way of summarising the data. We do this by **data reduction**.

> **Data reduction** gives a concise and accurate view of data. It shows the underlying patterns but does not swamp us with detail.

Data reduction has the advantages of;

● showing results in a compact form;

● highlighting overall patterns;

● using graphs or pictures;

● giving objective measures;

● allowing comparisons of different sets of data.

But, it has the disadvantages of:

● losing details of the original data;

● being irreversible.

After we have summarised a set of data, we have to present it in the best format. This is done by **data presentation**. So, to get reliable information:

● we start with data collection;

● summarise the results using data reduction;

● report the results using data presentation.

In practice, there is not such a clear distinction between the last two steps. In the rest of this chapter we will look at some diagrams for summarising and presenting data. The next chapter describes numerical summaries.

## In summary

**There is usually too much detail in raw data. Data reduction summarises the data and highlights the underlying patterns. Data presentation finds the best format for describing the results.**

# 4.2  Tables and frequency distributions

## 4.2.1 Using tables to present data

The easiest way of presenting numerical data is in a table. Table 4.1 shows the weekly sales data that was listed earlier.

**Table 4.1**

| Week | Quarter 1 | Quarter 2 | Quarter 3 | Quarter 4 | Total |
|------|-----------|-----------|-----------|-----------|-------|
| 1 | 51 | 84 | 49 | 30 | 214 |
| 2 | 60 | 91 | 44 | 32 | 227 |
| 3 | 58 | 82 | 41 | 30 | 211 |
| 4 | 56 | 78 | 45 | 32 | 211 |
| 5 | 62 | 76 | 38 | 31 | 207 |
| 6 | 69 | 75 | 28 | 29 | 201 |
| 7 | 58 | 66 | 37 | 30 | 191 |
| 8 | 76 | 57 | 40 | 41 | 214 |
| 9 | 80 | 78 | 42 | 45 | 245 |
| 10 | 82 | 65 | 22 | 44 | 213 |
| 11 | 68 | 50 | 25 | 47 | 190 |
| 12 | 90 | 61 | 26 | 53 | 230 |
| 13 | 72 | 54 | 21 | 54 | 201 |
| **Totals** | **882** | **917** | **458** | **498** | **2755** |

Now we have given the data some structure and can see some overall patterns. Demand is, for example, higher in the first two quarters and lower in the second two. But this table still gives only the raw data, and it is difficult to get a feel for a typical week's sales, or to find the highest sales, or the range. It would be useful to reduce the data some more and to emphasise the patterns. The smallest sales are 21, so we can start by counting the weeks with sales in the range, say, 20 to 29. Then we can count the number of observations in other ranges, as shown in Table 4.2. This gives a **frequency distribution**.

**Table 4.2**

| Range of sales | Number of weeks |
|---|---|
| 20 – 29 | 6 |
| 30 – 39 | 8 |
| 40 – 49 | 10 |
| 50 – 59 | 9 |
| 60 – 69 | 7 |
| 70 – 79 | 6 |
| 80 – 89 | 4 |
| 90 – 99 | 2 |

The 'ranges' are called **classes**, so we can talk about the 'class of 20 to 29', where 20 is the lower class limit, 29 is the upper class limit, and the class width is $29 - 20 = 9$. We can choose any convenient classes. The eight classes above might be too many, so we can reduce them to, say, the four shown in Table 4.3.

**Table 4.3**

| Range | Number of weeks |
|---|---|
| 20 – 39 | 14 |
| 40 – 59 | 19 |
| 60 – 79 | 13 |
| 80 – 99 | 6 |

The more we summarise data, the more detail we loose. So tables always need a compromise between being too long (when they give lots of details, but are complicated and hide underlying patterns) and being too short (when underlying patterns are clear, but most details are lost).

## In summary

**Tables are widely used for presenting numerical data. They give a simple format that can show a lot of information.**

## 4.2.2 Frequency distributions

The frequency distribution in Table 4.4 shows a company's weekly sales.

**Table 4.4**

| Class | Number of weeks |
|---|---|
| 20 – 39 | 14 |
| 40 – 59 | 19 |
| 60 – 79 | 13 |
| 80 – 99 | 6 |

There are six observations in the highest class, 80–99. Sometimes it is better to be a little less specific when defining classes, particularly if there is the odd outlying value. Suppose there is one observation of 120. It is better to add this to the highest class, rather than make an extra class some distance from the others. So we can define a class of '80 or more'. In the same way, it may be better to replace the precise '20 to 39' by the less precise '39 or fewer'.

We also have to be careful to define the boundaries between classes so that an observation can only fit into one class. We should not, for example, have classes of '20–30' and '30–40', as a value of 30 could be in either. Here we overcame this by using classes of '20–29' and '30–39'. But this only works if the data are discrete. With continuous data we have to be more careful. If we want to classify people by age, for example, we cannot use classes '20–29' and '30–39', as this leaves no place for people who are 29 $\frac{1}{2}$. So, we must describe the classes clearly and unambiguously, like '20 or more, but less than 30'.

# we Worked example 4.1

Last month Merryweather's Olde Shoppe paid its employees the following amounts:

202  457  310  176  480  277  87  391  325  120  554  94  362  221  274

145  240  437  404  398  361  144  429  216  282  153  470  303  338  209

Draw a frequency distribution of this data.

## Solution

First we have to choose the number of classes. Too few classes – say, three – would not show the overall patterns; too many classes – say, 20 – would be too detailed and confusing. A reasonable answer here is to use about six classes.

Now we have to define the class ranges. The range of wages is £87 to £554, so reasonable classes are:

less than £100,

£100 or more, but less than £200,

£200 or more, and less than £300, and so on

Notice that we are careful *not* to say 'more than £100 but less than £200' as someone earning exactly £100 would not appear in any class.

Now we can add the number of observations in each class to give the distribution, shown in Table 4.5.

**Table 4.5**

| Class | Frequency |
|---|---|
| Less than £100 | 2 |
| £100 or more, but less than £200 | 5 |
| £200 or more, but less than £300 | 8 |
| £300 or more, but less than £400 | 9 |
| £400 or more, but less than £500 | 5 |
| £500 or more | 1 |

Frequency distributions show the actual number of observations in each class. A useful extension is a **percentage frequency distribution** which shows the percentage of observations in each class. The percentage frequency distribution of the data in Worked Example 4.1 is shown in Table 4.6.

**Table 4.6**

| Class | Frequency | Percentage frequency |
|---|---|---|
| less than £100 | 2 | 6.67 |
| £100 or more, but less than £200 | 5 | 16.67 |
| £200 or more, but less than £300 | 8 | 26.67 |
| £300 or more, but less than £400 | 9 | 30.00 |
| £400 or more, but less than £500 | 5 | 16.67 |
| £500 or more | 1 | 3.33 |

Sometimes it is useful to look at the cumulative frequencies. Instead of showing the number of observations in a class, **cumulative frequency distributions** add all observations in lower classes. In Table 4.6 there were 2 observations in the first class, 5 in the second class and 8 in the third. So the cumulative frequency distribution shows 2 observations in the first class, $2 + 5 = 7$ in the second class, and $2 + 5 + 8 = 15$ in the third. Then we can extend this to give a **cumulative percentage frequency distribution**, as shown in the spreadsheet in Figure 4.1.

| | A | B | C | D | E |
|---|---|---|---|---|---|
| 1 | **Frequency distributions** | | | | |
| 2 | | | | | |
| 3 | **Class** | **Frequency** | **Cumulative frequency** | **Percentage frequency** | **Cumulative percentage frequency** |
| 4 | Less than £100 | 2 | 2 | 6.67 | 6.67 |
| 5 | £100 or more, but less than £200 | 5 | 7 | 16.67 | 23.33 |
| 6 | £200 or more, but less than £300 | 8 | 15 | 26.67 | 50.00 |
| 7 | £300 or more, but less than £400 | 9 | 24 | 30.00 | 80.00 |
| 8 | £400 or more, but less than £500 | 5 | 29 | 16.67 | 96.67 |
| 9 | £500 or more | 1 | 30 | 3.33 | 100.00 |

**Figure 4.1** Different types of frequency distribution.

# we Worked example 4.2

Draw a table of the frequency, cumulative frequency, percentage frequency and cumulative percentage frequency for the following discrete data.

150  141  158  147  132  153  176  162  180  165

174  133  129  119  103  188  190  165  157  146

161  130  122  169  159  152  173  148  154  171

## Solution

First, we have to define the classes. The data are discrete, so we can use $100-109$, $110-119$, $120-129$, and so on. Figure 4.2 shows the frequency distributions in a spreadsheet.

| | A | B | C | D | E |
|---|---|---|---|---|---|
| 1 | **Frequency distributions** | | | | |
| 2 | | | | | |
| 3 | **Class** | **Frequency** | **Cumulative frequency** | **Percentage frequency** | **Cumulative percentage frequency** |
| 4 | 100–109 | 1 | 1 | 3.33 | 3.33 |
| 5 | 110–119 | 1 | 2 | 3.33 | 6.67 |
| 6 | 120–129 | 2 | 4 | 6.67 | 13.33 |
| 7 | 130–139 | 3 | 7 | 10.00 | 23.33 |
| 8 | 140–149 | 4 | 11 | 13.33 | 36.67 |
| 9 | 150–159 | 7 | 18 | 23.33 | 60.00 |
| 10 | 160–169 | 5 | 23 | 16.67 | 76.67 |
| 11 | 170–179 | 4 | 27 | 13.33 | 90.00 |
| 12 | 180–189 | 2 | 29 | 6.67 | 96.67 |
| 13 | 190–199 | 1 | 30 | 3.33 | 100.00 |

**Figure 4.2** Frequency distribution for Worked Example 4.2.

## In summary

**Frequency distributions show the number of observations in different classes. We can extend this idea to show percentage frequency distributions and cumulative frequency distributions.**

# 4.3 Diagrams for presenting data

People are more likely to look at a diagram than to read text, so we can use diagrams to make data presentation more interesting – remember that one picture is worth a thousand words. There are several types of diagram for summarising data, including:

- graphs
- pie charts
- bar charts
- pictograms
- histograms

## 4.3.1 Drawing graphs

Graphs use rectangular – or Cartesian – axes to show the relationship between two variables. The horizontal axis is called the $x$-axis and it shows the **independent variable** – which is the one that we can control. The vertical axis is called the $y$-axis and it shows the **dependent variable**, whose value is set by $x$. So $x$ might be the amount spent on advertising and $y$ is the resulting sales; $x$ might be the interest rate for loans and $y$ is the amount borrowed from banks; $x$ might be the price charged for a service and $y$ the resulting demand.

  The point where the two axes cross is the **origin**, and is the point where both $x$ and $y$ have the value zero. Then $y$ is positive for all points above the origin, and negative for all points below; $x$ is positive for all points to the right of the origin, and negative for all points to the left (see Figure 4.3).

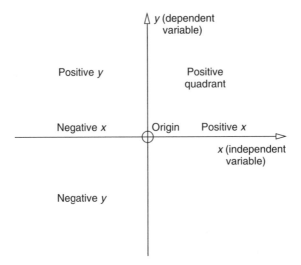

**Figure 4.3**   Cartesian co-ordinates.

We can describe any point on a graph by two numbers called **co-ordinates**. The first number gives the distance along the $x$-axis from the origin, and the second number gives the distance up or down the $y$-axis. The point $x = 3$, $y = 4$, for example, is 3 units along the $x$-axis and 4 units up the $y$-axis. Co-ordinates are usually written as $(x, y)$, so we can describe this point as $(3, 4)$. The only thing you have to be careful about is that $(3, 4)$ is not the same as $(4, 3)$ or $(-3, 4)$, as shown in Figure 4.4.

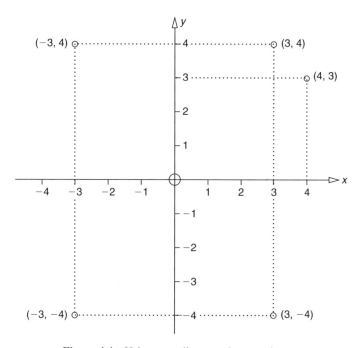

**Figure 4.4**   Using co-ordinates to locate points.

Often we are only interested in positive values – with a graph of income against sales, for example – so we only show the positive quadrant.

## we Worked example 4.3

Plot the following points on a graph.

| $x$ | 2 | 5 | 7 | 10 | 15 | 20 |
|---|---|---|---|---|---|---|
| $y$ | 7 | 20 | 22 | 29 | 47 | 59 |

## Solution

We only have positive numbers, so we only need draw the positive quadrant of the graph. Then the first point, $(2, 7)$, is 2 units along the $x$-axis and 7 units up the $y$-axis. This is shown as point $A$ in Figure 4.5. The second point, $(5, 20)$, is 5 units along the $x$-axis and 20 units up the $y$-axis, and is shown by point B. We can add the other points in the same way.

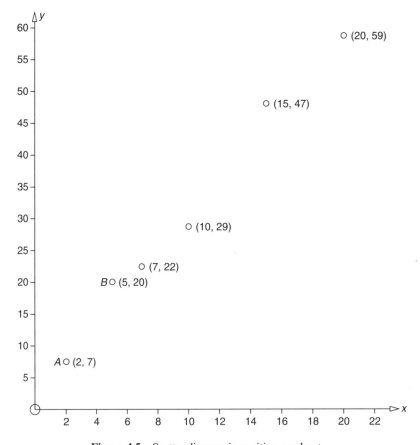

**Figure 4.5** Scatter diagram in positive quadrant.

Figure 4.5 is a **scatter diagram**, which shows the individual points. If we draw a scatter diagram of the sales figures given in section 4.1 – with the week as the independent variable and sales as the dependent variable – we get the result shown in Figure 4.6.

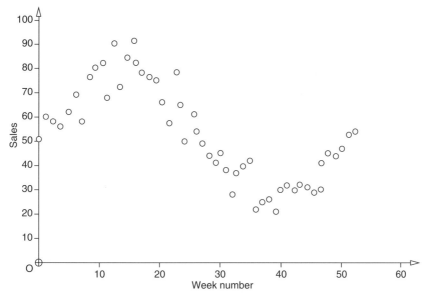

**Figure 4.6**   Scatter diagram of sales against week.

The overall patterns are often clearer if we join the points, as shown in Figure 4.7. The sales are highest around week 12 and lowest around week 39. But there are small variations and the graph is not smooth. We are usually more interested in the overall pattern than the small variations, so we can draw a smooth trend line through the points, as shown in Figure 4.8.

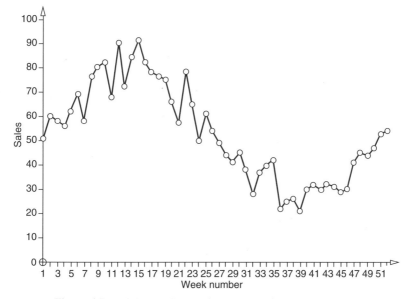

**Figure 4.7**   Joining popints to give a graph of sales against week.

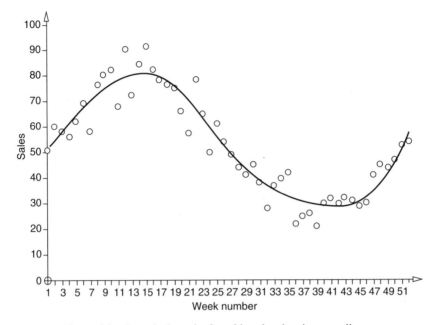

**Figure 4.8**   Smoothed graph of weekly sales showing overall pattern.

Sometimes it is difficult to choose a scale for the y-axis. Nevertheless, the choice is important as the scale affects the overall picture. At first sight, Figure 4.9 shows a stable pattern with small variations around a low constant value, while Figure 4.10 shows widely varying values, which are high in the first half, and then low in the second half. If you look more carefully, however, you will see that these two graphs show the same data as Figure 4.6 but with different scales for the y-axis.

Choosing the scale for the y-axis is often a matter of personal judgement, but some suggestions can be made:

● always label the axes clearly and accurately;

● show the scale on both axes;

● the highest value on the scale should be slightly more than the largest observation;

● wherever possible, scales on the axes should start at zero;

● if the scale does not start at zero, you must show this clearly – perhaps putting marks through the axis to show a break;

● where appropriate, give the source of the data;

● where appropriate, give the graph a title.

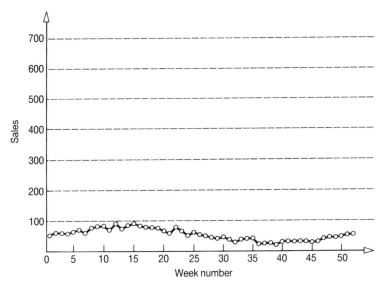

**Figure 4.9** Graph of apparently stable sales.

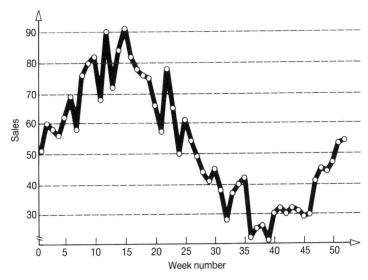

**Figure 4.10** Graph of apparently variable sales.

We can compare sets of data by plotting several graphs on the same axes. Figure 4.11, for example, shows how the price of a basic commodity has varied each month over the past five years. The price differences are small, so we have emphasised them by using a wide scale for the y-axis.

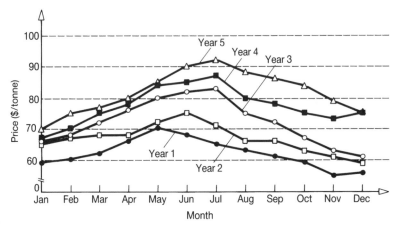

**Figure 4.11** Price in $ per tonne of commodity by month. (Source: UN Digest)

## In summary

**Graphs use rectangular co-ordinates to show a relationship between two variables: an independent one, $x$, and a dependent one, $y$.**

### 4.3.2 Straight-line graphs

There are many standard patterns in graphs. One particularly important pattern shows a linear relationship – so the trend line through the points it is a straight line.

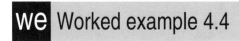 **Worked example 4.4**

Table 4.7 shows a company's quarterly profit and the corresponding share price on the London Stock Exchange. Draw a graph of these data.

**Table 4.7**

|  | Year 1 | | | | Year 2 | | | | Year 3 | | | |
|---|---|---|---|---|---|---|---|---|---|---|---|---|
|  | Q1 | Q2 | Q3 | Q4 | Q1 | Q2 | Q3 | Q4 | Q1 | Q2 | Q3 | Q4 |
| Profit(£m) | 12.1 | 12.2 | 11.6 | 10.8 | 13.0 | 13.6 | 11.9 | 11.7 | 14.2 | 14.5 | 12.5 | 13.0 |
| Share price (p) | 122 | 129 | 89 | 92 | 12 | 135 | 101 | 104 | 154 | 156 | 125 | 136 |

Sources: company reports and the *Financial Times*

# Solution

The independent variable is the one that is responsible for changes, which is the company profit. The dependent variable is the one we are trying to explain, which is the share price. Figure 4.12 shows a graph of these figures, with the scales chosen to emphasise the linear relationship between profit and share price.

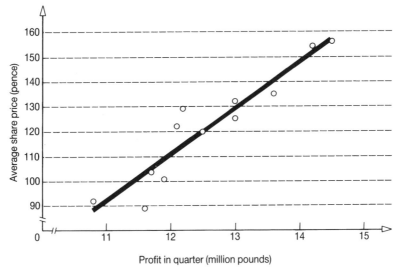

**Figure 4.12**   Graph of share price against sales of Worked Example 4.4.

We can describe a straight line graph by a simple equation.

> The general equation of a straight line is:
>
> $$y = ax + b$$

Where:

$x$ is the independent variable

$y$ is the dependent variable

$a$ and b are constants

We might find, for example, that $y = 4x + 10$. Then for every value of the independent variable $x$ there is a corresponding value of the dependent variable $y$.

## we Worked example 4.5

Draw a graph of $y = 10x + 50$.

## Solution

This is a straight-line graph of the standard form $y = ax + b$, with $a = 10$ and $b = 50$. We only need to find two points in order to draw a straight-line graph. Here we can take two convenient points, say $x = 0$ and $x = 20$. Then:

when $x = 0$:  $y = 10x + 50 = (10 \times 0) + 50 = 50$, which gives the point $(0, 50)$

when $x = 20$:  $y = 10x + 50 = (10 \times 20) + 50 = 250$, which gives the point $(20, 250)$

Plotting these points and then connecting them gives the graph shown in Figure 4.13.

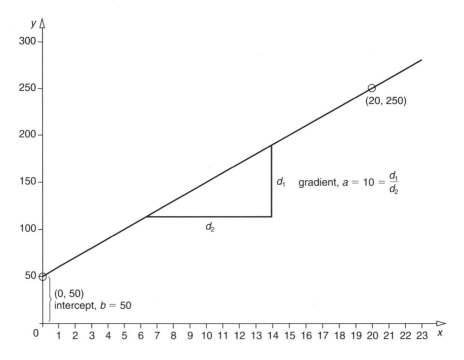

**Figure 4.13**  Graph of $y = 10x + 50$ for Worked Example 4.5.

If you look at the graph in Figure 4.13, there are two obvious features:

- the **intercept**, which shows where the line crosses the $y$-axis, and

- the **gradient**, which shows how steep the line is.

A line crosses the $y$-axis when $x$ has the value 0, so the intercept is the value of $y$ when $x = 0$. If we substitute $x = 0$ into the general equation $y = ax + b$, we find:

$$y = (a \times 0) + b$$

or,

$$y = b$$

In other words, $b$ is the intercept of the line.

The gradient shows how quickly the line is rising (or falling), and we can define it as the increase (or decrease) in $y$ for a unit increase in $x$. Taking the graph of $y = 10x + 50$, we can find the change in $y$ when $x$ rises between, say, 10 and 11. (These are just two convenient points as a straight line has the same gradient at any point.) Then:

when $x = 10$:  $y = 10x + 50 = (10 \times 10) + 50 = 150$

when $x = 11$:  $y = 10x + 50 = (10 \times 11) + 50 = 160$

So the gradient is $(160 - 150) = 10$. But you can see that this is the value of $a$ in the initial equation $-y = 10x + 50$. This is no coincidence, and the gradient of a straight line is always given by the value of $a$ in the equation $y = ax + b$.

# we Worked example 4.6

Describe the graph of the equation $y = 100 - 5x$.

## Solution

This is a straight-line graph of the form $y = ax + b$ with $a = -5$ and $b = 100$. We only need to find two points to draw the line, so we can take two convenient ones, such as $x = 0$ and $x = 10$. Then:

when $x = 0$:  $y = 100 - 5x = 100 - (5 \times 0) = 100$, giving the point $(0, 100)$

when $x = 10$:  $y = 100 - 5x = 100 - (5 \times 10) = 50$, giving the point $(10, 50)$

Plotting these two points and drawing the line between them gives the result shown in Figure 4.14.

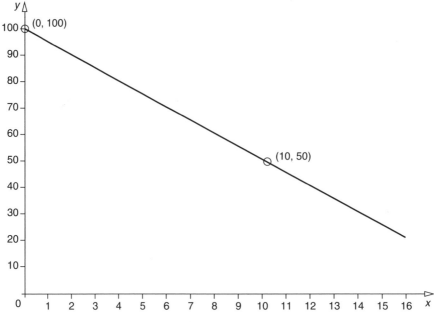

**Figure 4.14** Graph of $y = 100 - 5x$ for Worked Example 4.6.

In this example the intercept is 100 and the gradient is -5 (showing that $y$ decreases as $x$ increases).

## In summary

**Straight-line graphs have the form $y = ax + b$, where $a$ is the gradient and $b$ is the intercept.**

### 4.3.3 Drawing graphs of other equations

We can draw a straight-line graph using any two points, but we need more points to draw more complicated functions. The easiest way to draw a graph of a complex function is to take a series of convenient values for the independent variable $x$, and substitute these into the equation to find corresponding values for the dependent variable $y$. This gives a series co-ordinates $(x, y)$. Then we can join these points to show the graph of the relationship.

## we Worked example 4.7

Draw the graph of the quadratic equation $y = 2x^2 + 3x - 3$, between $x = -6$ and $x = 5$.

## Solution

Any equation like this, with the general form $y = ax^2 + bx + c$, is called a **quadratic equation**. The shape of this graph is not obvious from the equation, but if we take convenient values of $x$, we can substitute into the equation to get a series of co-ordinates. For example:

when $x = -6$:  $y = 2x^2 + 3x - 3 = 2 \times (-6)^2 + 3 \times (-6) - 3 = 51$

when $x = -5$:  $y = 2x^2 + 3x - 3 = 2 \times (-5)^2 + 3 \times (-5) - 3 = 32$

and so on, to give the following table.

| $x$ | −6 | −5 | −4 | −3 | −2 | −1 | 0 | 1 | 2 | 3 | 4 | 5 |
|---|---|---|---|---|---|---|---|---|---|---|---|---|
| $y$ | 51 | 32 | 17 | 6 | −1 | −4 | −3 | 2 | 11 | 24 | 41 | 62 |

Plotting these points on rectangular axes and joining them together gives the graph in Figure 4.15.

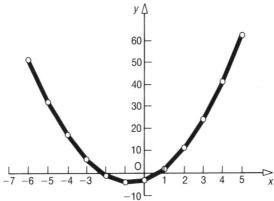

**Figure 4.15**   Graph of $y = 2x^2 + 3x - 3$ for Worked Example 4.7.

Quadratic equations always have a U-shaped curve when the $x^2$ term is positive. When the $x^2$ term is negative the graph is upside-down, so that it looks like a hill rather than a valley.

We can use this method of drawing graphs for any other function. In Worked Example 4.8 we look at an exponential function. This is based on the exponential constant, e, which is defined as $e = 2.7182818....$ Although this seems a strange number, there are good theoretical reasons for using it.

## we Worked example 4.8

Draw a graph of $y = e^x$ for values of $x$ between 0 and 10.

## Solution

Taking a range of values for $x$ and substituting these into the equation for $y$, gives the following table of values.

| $x$ | 0 | 1 | 2 | 3 | 4 | 5 | 6 | 7 | 8 | 9 | 10 |
|---|---|---|---|---|---|---|---|---|---|---|---|
| $e^x$ | 1 | 2.7 | 7.4 | 20.1 | 54.6 | 148.4 | 403.4 | 1096.6 | 2981.0 | 8103.1 | 22026.5 |

We can draw these on a graph to show the characteristic exponential growth curve shown in Figure 4.16.

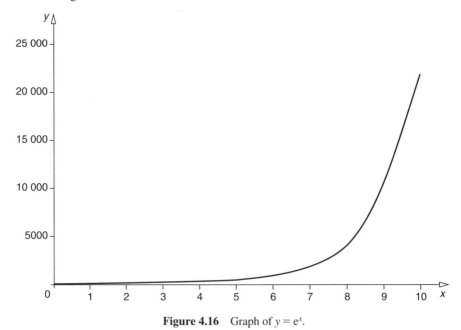

**Figure 4.16**  Graph of $y = e^x$.

## In summary

**Graphs can show a picture of any function. The easiest way of drawing these is to take a range of values for $x$ and substitute these into the equation to find corresponding values for $y$. This gives a series of co-ordinates $(x, y)$ which we can plot and join with a line.**

### 4.3.4 Pie charts

Pie charts are simple diagrams that show the numbers of observations in different categories. To draw a pie chart, we divide the data into distinct categories and count the number of observations in each. Then sectors of a circle – or slices of a pie – represent each category. The angle at the centre of the slice – and hence the area of the slice – is proportional to the number of observations in the category.

## we Worked example 4.9

Sales in four regions are given in Table 4.8. Draw a pie chart for these.

**Table 4.8**

| Region | Sales |
|--------|-------|
| North  | 25    |
| South  | 10    |
| East   | 45    |
| West   | 25    |
| **Total** | **100** |

## Solution

There are $360°$ in a circle, and these represent 100 observations. So each observation is represented by an angle of $360/100 = 3.6°$ at the centre of the circle. Sales in the North region are represented by a slice with an angle of $25 \times 3.6 = 90°$ at the centre; sales in the South region are represented by a slice with an angle of $10 \times 3.6 = 36°$, and so on. These drawings are easily done by computer, and Figure 4.17(a) shows a typical result.

We can improve the appearance of this basic chart in several ways: Figure 4.17(b) shows the same results with slices sorted into order, percentages calculated and a three-dimensional effect added.

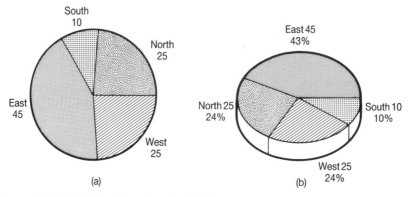

**Figure 4.17** (a) Basic pie chart for Worked Example 4.9; (b) improved version same chart.

We can only use pie charts to show a small number of categories. When there are more than, say, six categories they become too complicated and lose their impact.

## In summary

**Pie charts use sectors of a circle to show the relative frequency of observations. Although thay have an immediate impact, they are only useful with a few categories of data.**

### 4.3.5 Bar charts

Bar charts also show the number of observations in different categories of data, but this time the numbers of observations are shown as lines or bars. Each bar represents a category of data, and the length of the bar is proportional to the number of observations.

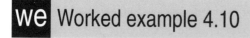

Draw a bar chart of the regional sales in Worked Example 4.9.

## Solution

Figure 4.18 shows a straightforward bar chart for these data.

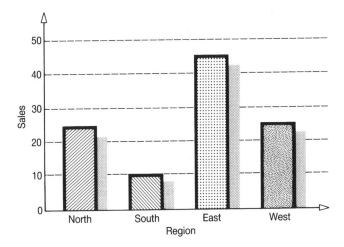

**Figure 4.18**    Bar chart for Worked Example 4.10.

There are several variations of the basic bar chart, and you have to choose the best one for any particular presentation. Remember, though, that the purpose of diagrams is to present a summary of the data and is not necessarily to draw the prettiest picture.

## we Worked example 4.11

A health district classifies the number of beds in five hospitals as shown in Table 4.9. Draw a bar chart of these data.

**Table 4.9**

|  | Hospital | | | | |
|---|---|---|---|---|---|
|  | Foothills | General | Southern | Heathview | St John |
| Maternity | 24 | 38 | 6 | 0 | 0 |
| Surgical | 86 | 85 | 45 | 30 | 24 |
| Medical | 82 | 55 | 30 | 30 | 35 |
| Psychiatric | 25 | 22 | 30 | 65 | 76 |

## Solution

We can draw several bar charts of these data. Figure 4.19(a) shows a vertical format that emphasises the number of beds of each type. If we want to highlight the sizes of the hospitals, we can 'stack' the bars to give the single bars shown in Figure 4.19(b). We can also draw the percentages shown in Figure 4.19(c).

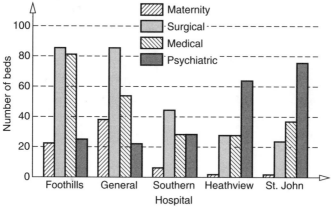

**Figure 4.19 (a)**   Basic bar chart of hospital beds for Worked Example 4.11.

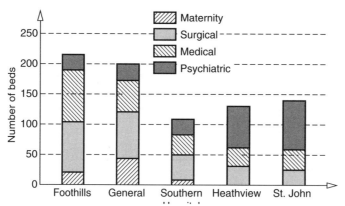

**Figure 4.19 (b)**   Emphasising the number of beds in each hospital.

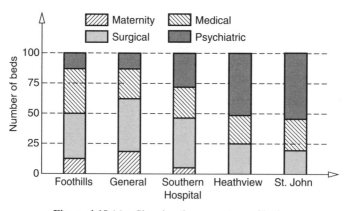

**Figure 4.19 (c)**   Showing the percentage of beds.

# In summary

**Bar charts use the length of bars to represent the numbers of observations in different categories of data.**

## 4.3.6 Pictograms

Pictograms are similar to bar charts, except that the bars are replaced by sketches of the things being described. We could show the percentage of people owning cars, for example, by the rows of cars in Figure 4.20. Here, each 10% of people is represented by one car.

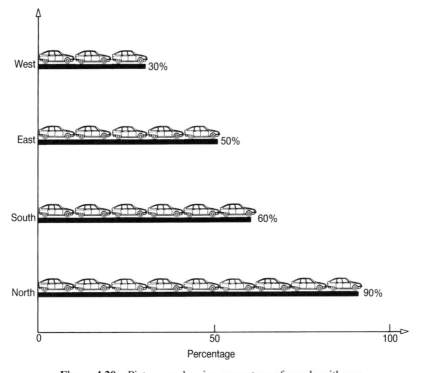

**Figure 4.20**   Pictogram showing percentage of people with cars.

Pictograms are eye-catching and effective in giving general impressions, but there are two things to notice:

- Fractional values do not work well. If 53% of people owned cars in Figure 4.20 we would have a row of 5.3 cars – and the 0.3 of a car may be confusing.

- Pictograms must show different numbers of observations by different numbers of sketches. You cannot draw a single big sketch, like Figure 4.21. It is the length of the rows that is important, but with single sketches it is the area that has the impact – and doubling the number of observations makes a single sketch twice as long, but with four times the area.

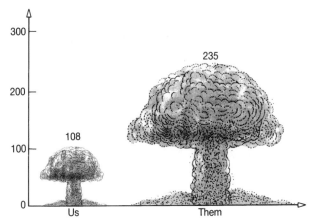

**Figure 4.21**   Poor pictogram showing the number of nuclear missiles.

## In summary

**Pictograms replace the bars in bar charts by sketches. These can attract attention, but the results are not very accurate and need careful interpretation.**

### 4.3.7 Histograms

**Histograms** look like vertical bar charts, but there are some important differences. They show frequency distributions for continuous data, so the *x*-axis has a continuous scale. This means that the width of the bars, as well as their height, is important. With bar charts only the height of the bar is important, but in histograms it is both the width and the height – or the area. You can see this in Figure 4.22 which is a histogram of wages from data in Worked Example 4.1.

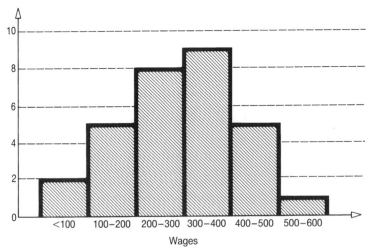

**Figure 4.22**   Histogram of wages.

Each class is the same width, so the areas are set by the height of the bars. In effect, this is the same as a bar chart. However, if the classes have different widths we have to be much more careful.

## we Worked example 4.12

Draw a histogram of the data in Table 4.10.

**Table 4.10**

| Class | Frequency |
|---|---|
| Less than 10 | 8 |
| 10 or more, but less than 20 | 10 |
| 20 or more, but less than 30 | 16 |
| 30 or more, but less than 40 | 15 |
| 40 or more, but less than 50 | 11 |
| 50 or more, but less than 60 | 4 |
| 60 or more, but less than 70 | 2 |
| 70 or more, but less than 80 | 1 |
| 80 or more, but less than 90 | 1 |

## Solution

Figure 4.23 shows a histogram using the classes given.

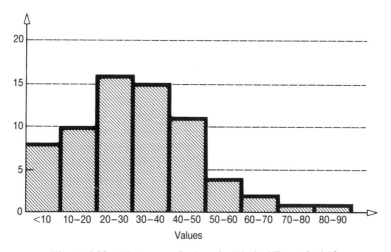

**Figure 4.23**   Histogram of values for Worked Example 4.12.

This diagram has only eight observations in the last four classes, so we might be tempted to combine these into one class, and draw the histogram in Figure 4.24(a). But this is wrong. The area shows the number of observations, and a class that is four times as wide as the other classes, and eight units high contains $4 \times 8 = 32$ observations. To show only eight observations, the bar should only be two units high, as shown in Figure 4.24(b).

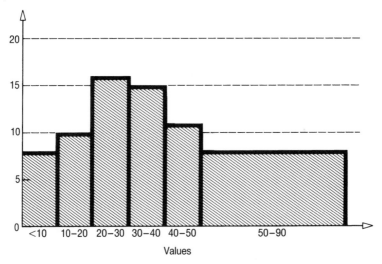

**Figure 4.24 (a)**   Incorrect histogram combining the last four classes.

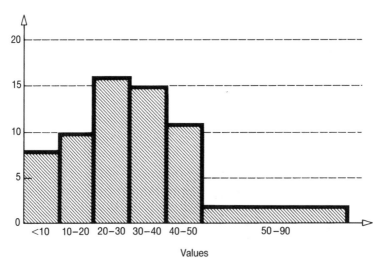

**Figure 4.24**   Correct histogram.

Both the width and the height of the bars are important in histograms, so their shape depends on the way we define the classes. This shows the major disadvantage that relatively small changes to the width of classes can give major changes to the shape of the bars.

There is another problem with open-ended classes. How, for example, can we deal with a class of values 'greater than 20'. The only answer is to avoid this kind of definition, and assume that 'greater than 20' really means 'greater than 20 and less than 22'.

## In summary

**Histograms show frequency distributions for continuous data. The frequencies are shown by the areas of the bars. Because it can be difficult to draw a reasonable histogram, it is often better to use bar charts.**

## Chapter review

After collecting data, we must process them to give useful information. The processing involves data reduction and presentation. This chapter showed how to summarise data in diagrams. In particular, it:

- showed why data reduction is needed to give useful information;
- outlined the steps in data reduction and presentation;
- designed tables of numerical data;
- described frequency distributions;
- drew graphs to show relationships between variables;
- drew pie charts, bar charts and pictograms to show relative frequencies;
- drew histograms for continuous data.

## Problems

4.1 A survey question gets the answer Yes from 47% of men and 38% of women, No from 32% of men and 53% of women, and Don't know from the rest. How would you present this result?

4.2 A group of people earned the following average wages:

| 221 | 254 | 83 | 320 | 367 | 450 | 292 | 161 | 216 | 410 | 380 | 355 | 502 | 144 | 362 |
| 112 | 387 | 324 | 576 | 156 | 295 | 77 | 391 | 324 | 126 | 154 | 94 | 350 | 239 | 263 |
| 276 | 232 | 467 | 413 | 472 | 361 | 132 | 429 | 310 | 272 | 408 | 480 | 253 | 338 | 217 |

Draw a frequency distribution, histogram, percentage frequency and cumulative frequency distribution of this data.

4.3 The following data have been collected by a company. Put them into a spreadsheet and summarise the results.

| 245 | 487 | 123 | 012 | 159 | 751 | 222 | 035 | 487 | 655 | 197 | 655 | 458 | 766 | 123 | 453 | 493 | 444 | 123 | 537 |
| 254 | 514 | 324 | 215 | 367 | 557 | 330 | 204 | 506 | 804 | 941 | 354 | 226 | 870 | 652 | 458 | 425 | 248 | 560 | 510 |
| 234 | 542 | 671 | 874 | 710 | 702 | 701 | 540 | 360 | 654 | 323 | 410 | 405 | 531 | 489 | 695 | 409 | 375 | 521 | 624 |
| 357 | 678 | 809 | 901 | 567 | 481 | 246 | 027 | 310 | 679 | 548 | 227 | 150 | 600 | 845 | 521 | 777 | 304 | 286 | 220 |
| 667 | 111 | 485 | 266 | 472 | 700 | 705 | 466 | 591 | 398 | 367 | 331 | 458 | 466 | 571 | 489 | 257 | 100 | 874 | 577 |

4.4    The number of students taking a course in the past ten years is summarised in Table 4.11. How would you present these data in diagrams?

**Table 4.11**

| Year   | 1  | 2  | 3  | 4  | 5  | 6  | 7  | 8  | 9  | 10 |
|--------|----|----|----|----|----|----|----|----|----|----|
| Male   | 21 | 22 | 20 | 18 | 28 | 26 | 29 | 30 | 32 | 29 |
| Female | 4  | 6  | 3  | 5  | 12 | 16 | 14 | 19 | 17 | 25 |

4.5    Table 4.12 shows a company's quarterly profit and its share price on the London Stock Exchange. How would you present these data?

**Table 4.12**

|                   | Year 1 | | | | Year 2 | | | | Year3 | | | |
|-------------------|-----|-----|-----|-----|-----|-----|-----|-----|-----|-----|-----|-----|
| Quarter           | Q1  | Q2  | Q3  | Q4  | Q1  | Q2  | Q3  | Q4  | Q1  | Q2  | Q3  | Q4  |
| Profit (£m)       | 36  | 45  | 56  | 55  | 48  | 55  | 62  | 68  | 65  | 65  | 69  | 74  |
| Share price (p)   | 137 | 145 | 160 | 162 | 160 | 163 | 166 | 172 | 165 | 170 | 175 | 182 |

Sources: company reports and the *Financial Times*

4.6    Four regions of Yorkshire classify companies according to primary, manufacturing, transport, retail and service. The number of companies working in each region in each category is shown in Table 4.13. How could you present these data?

**Table 4.13**

|           | Industry type | | | | |
|-----------|---------|---------------|-----------|--------|---------|
|           | Primary | Manufacturing | Transport | Retail | Service |
| Daleside  | 143     | 38            | 10        | 87     | 46      |
| Twendale  | 134     | 89            | 15        | 73     | 39      |
| Underhill | 72      | 67            | 11        | 165    | 55      |
| Perithorp | 54      | 41            | 23        | 287    | 89      |

4.7    Draw a graph of:

(a)  $y = 10$

(b)  $y = x + 10$

(c)  $y = x^2 + x + 10$

(d)  $y = x^3 + x^2 + x + 10$

4.8     Draw a histogram of the data in Table 4.14. How could you combine the last (a) two classes, (b) three classes?

**Table 4.14**

| Class | Frequency |
|-------|-----------|
| Less than 100 | 120 |
| 100 or more, but less than 200 | 185 |
| 200 or more, but less than 300 | 285 |
| 300 or more, but less than 400 | 260 |
| 400 or more, but less than 500 | 205 |
| 500 or more, but less than 600 | 150 |
| 600 or more, but less than 700 | 75 |
| 700 or more, but less than 800 | 35 |
| 800 or more, but less than 900 | 15 |

# CS Case study – High Acclaim Importers

The finance director of High Acclaim Importers was giving a summary of business to a group of shareholders. He asked Jim Bowlers to collect some data from company records for his presentation. He wanted some concise figures that would look good on overhead slides.

At first Jim had been worried by the amount of detail available. The company seemed to have enormous amounts of data, ranging from transaction records in a computerised database, to subjective management views that were never written down.

Jim did a conscientious job of collecting data and he looked pleased as he approached the finance director. As he handed over the results – which are shown in Table 4.15 – Jim explained: 'Some of our most important trading results are shown in this table. We trade in four regions, so for movements between each of these I have recorded seven key facts. The following table shows the number of units shipped (in hundreds), the average income per unit (in pounds sterling), the percentage gross profit, the percentage return on investment, a measure (between 1 and 5) of trading difficulty, the number of finance administrators employed in each area, and the number of agents.'

**Table 4.15**

| | | To | | | |
|------|--------|--------|---------|-------|--------|
| | | Africa | America | Asia | Europe |
| From | Africa | 105, 45, 12, 4, 4, 15, 4 | 85, 75, 14, 7, 3, 20, 3 | 25, 60, 15, 8, 3, 12, 2 | 160, 80, 13, 7, 2, 25, 4 |
| | America | 45, 75, 12, 3, 4, 15, 3 | 255, 120, 15, 9, 1, 45, 5 | 60, 95, 8, 2, 2, 35, 6 | 345, 115, 10, 7, 1, 65, 5 |
| | Asia | 85, 70, 8, 4, 5, 20, 4 | 334, 145, 10, 5, 2, 55, 6 | 265, 85, 8, 3, 2, 65, 7 | 405, 125, 8, 3, 2, 70, 8 |
| | Europe | 100, 80, 10, 5, 4, 30, 3 | 425, 120, 12, 8, 1, 70, 7 | 380, 105, 9, 4, 2, 45, 5 | 555, 140, 10, 6, 110, 8 |

The finance director looked at the figures for a few minutes and then asked for some details on how trade had changed over the past ten years. Jim replied that in the past ten years the volume of trade had risen by 1.5, 3, 2.5, 2.5, 1, 1, 2.5, 3.5, 3 and 2.5 per cent respectively, while the average price had risen by 4, 4.5, 5.5, 7, 3.5, 4.5, 6, 5.5, 5 and 5 per cent respectively.

The finance director looked up from his figures and said: 'I was hoping for something a bit briefer and with a bit more impact. Could you give me the figures in a revised format by this afternoon?'

How would you help Jim Bowlers to put the figures into a better format for the presentation?

# 5

# Using numbers to describe data

## Contents

## Chapter outline

This chapter looks at some numerical descriptions of data. After reading the chapter you should be able to:

- understand the need to measure data
- calculate arithmetic means, medians and modes
- discuss the benefits and drawbacks of these measures
- calculate ranges and quartile deviations
- calculate mean absolute deviations, variances and standard deviations
- use indices to describe changes

# 5.1 Measures for business data

The previous chapter showed how to reduce data and present them in diagrams. These diagrams are good at giving overall impressions. A pictogram, for example, can have a visual impact, but it uses no numerical measures. This chapter shows how we can measure data.

Suppose we have the following set of data:

32 33 36 38 37 35 35 34 33 35 34 36 35 34 36 35 37 34 35 33

How can we measure the data and compare it with the following set?

2 8 21 10 17 24 18 12 1 16 12 3 7 8 9 10 9 21 19 6

We can start by drawing frequency distributions. Figure 5.1 shows how a statistical package called Minitab describes the two sets of data in 'dotplots'.

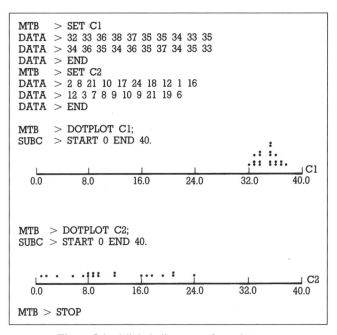

**Figure 5.1**   Minitab diagrams of two data sets.

There are twenty observations in each set of data, but there are two clear differences:

● values in the first set of data are higher than the second set, with the centre around 35 rather than 12;

● the second set of data is more spread out than the first.

So, we can start looking for two measures of data:

- A **measure of location** to show the centre of the data.
- A **measure of spread** to show how widely the data are spread out.

These two measures are shown in the histogram in Figure 5.2.

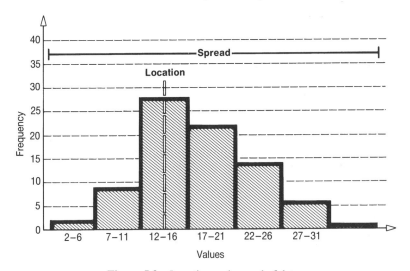

**Figure 5.2**    Location and spread of data.

## In summary

**Diagrams give overall impressions of data but they do not give objective measures. We need numerical measures to describe data more accurately. Two measures give the location and spread of data.**

# 5.2  Finding the average

## 5.2.1  Introduction

Most people are familiar with the notion of an average as some sort of typical value. If we know that the average age of students in a night class is 45, this gives some idea of what the class looks like; if the average income of a group of people is £60 000 a year, we know that they are prosperous; if houses in a village have an average of six bedrooms we know they are large; and so on.

Unfortunately, if we take a simple average the results can be quite misleading. The group with an average income of £60 000 a year might be ten people, nine with an income of £10000 a year and one with an income of £510000 a year. The village where houses have an average of six bedrooms might have 99 houses with two bedrooms each, and a stately home with 402 bedrooms. In both of these cases the quoted average is accurate, but it does not give a typical value. To get around this problem, we can define different types of average.

- The **mean** gives the centre of gravity of the data.

- The **median** is the middle value.

- The **mode** is the most frequent value.

Each of these is useful in different circumstances, as we shall see in the following sections.

## In summary

**There are several measures for the location of data. The most useful are the mean, median and mode.**

## 5.2.2 Mean

If you ask a group of people to find the average of 2, 4 and 6 is, they will usually say 4. To get this, they add all observations together and then divide the total by the number of observations, so $(2 + 4 + 6)/3 = 12/3 = 4$. This is the **arithmetic mean**, which is the most widely used measure of location. It is usually abbreviated to the **mean**, and is sometimes called the arithmetic average, simple average, or average.

When we have to add a lot of values, the easiest way of describing the arithmetic is to use subscripts. Suppose you have a series of observations. We can call this set of data $x$, and identify each observation by a subscript. Then $x_1$ is the first observation, $x_2$ is the second, $x_3$ is the third and $x_n$ is the $n$th observation.

Using this notation we can refer to a general observation as $x_i$. Then when $i = 5$, $x_i$ is $x_5$. At first sight this may not seem very useful, but in practice it saves a lot of effort. Suppose, for example, we want to find $x_1 + x_2 + x_3 + x_4$. Using the subscripts we can write this as:

sum of $x_i$ when $i = 1, 2, 3, 4$

Better still, there is a standard abbreviation for 'the sum of', which is the Greek capital letter sigma, $\Sigma$. Then we get:

$\Sigma x_i$ when $i = 1, 2, 3, 4$

Ordinarily, the range of $i$ that we want is written above and below the $\Sigma$ to give:

$$\sum_{i=1}^{4} x_i$$

The variable $i$ is always assumed to increase in steps of 1.

## we Worked example 5.1

(a) If you have a set of observations, $x$, how would you describe the sum of the first 10 observations?

(b) If you have a set of observations, $a$, how would you describe the sum of observations numbered 18 to 35?

## Solution

(a) We want the sum of $x_i$ when $i = 1$ to 10. We can write this as:

$$\sum_{i=1}^{10} x_i$$

(b) Now we want the sum of $a_i$ from $i = 18$ to 35. This is:

$$\sum_{i=18}^{35} a_i$$

We can use this notation to define the mean of a set of $n$ observations, $x_i$. This mean is usually called $\bar{x}$, which is expressed '$x$ bar'.

$$\mathbf{mean} = \bar{x} = \frac{x_1 + x_2 + x_3 + \ldots + x_n}{n} = \frac{\sum_{i=1}^{n} x_i}{n}$$

## we Worked example 5.2

Anne Birch recorded the times she needed to inspect units on a production line as 3, 4, 1, 7 and 1 minutes. What is the mean?

## Solution

To find the mean we add the observations, $x_i$, and divide by the number of observations, $n$:

$$\text{mean} = \frac{\sum x_i}{n} = \frac{3+4+1+7+1}{5} = \frac{16}{5} = 3.2 \text{ minutes}$$

Notice that in Worked Example 5.2 we used the abbreviation $\Sigma x$ for the summation. When we want to add all of a set of data, and there can be no misunderstanding, we usually replace the rather cumbersome $\sum_{i=1}^{n} x_i$ by the simpler, $\Sigma x$.

The mean of a set of integers is often a fraction. The average number of children in a family, for example, is 1.8. So the mean gives a reasonable measure of location, but it does not give a typical value.

Sometimes the mean does not even give a reasonable value. Suppose a company has three owner/directors who vote on the percentage of profits to be kept for future investment. If the directors suggested 5%, 7% and 9%, then the mean is 7%. But if the directors own 10, 10 and 1000 shares respectively, the views of the third director should carry more weight than the others. Then we can define a **weighted mean** where each observation is given a different weight.

$$\text{weighted mean} = \frac{\Sigma w_i x_i}{\Sigma w_i}$$

where:

$x_i$ = observation $i$

$w_i$ = weight given to observation $i$

With the owner/directors we could weight their views by the number of shares held to give a weighted mean:

$$\text{weighted mean} = \frac{(10 \times 5) + (10 \times 7) + (1000 \times 9)}{10 + 10 + 1000} = 8.94$$

We can extend this idea of weighted means to find the mean of data that has already had some processing. Suppose you have a set of data that has already been summarised in a frequency table. As we do not know the actual observations, we cannot find the true mean. However, we know the number of observations in each class, and can find an approximate value. For this, we assume that all observations in a class lie at its midpoint. If, for example, we have ten observations in a class $20-29$, we assume that all ten observations have the value $(20 + 29)/2 = 24.5$. Then if we have a frequency distribution for $n$ observations, with $f_i$ as the number of observations in class $i$ and $x_i$ as the midpoint of class $i$, the sum of all observations is $\Sigma f_i x_i$. This is usually abbreviated to $\Sigma fx$, so:

**For grouped data:**

$$\text{mean} = \bar{x} = \frac{\Sigma fx}{\Sigma f} = \frac{\Sigma fx}{n}$$

As you would expect, the top line in this equation gives the sum of all observations, while the bottom line is the number of observations.

## we Worked example 5.3

Find the mean of the following continuous frequency distribution.

| Class | 0-0.99 | 1.00–1.99 | 2.00–2.99 | 3.00–3.99 | 4.00–4.99 | 5.00–5.99 |
|---|---|---|---|---|---|---|
| Frequency | 1 | 4 | 8 | 6 | 3 | 1 |

## Solution

The values of $x_i$ are the midpoints of the classes. The midpoint of the first class, $x_1$, is $(0+0.99)/2 = 0.5$; the midpoint of the second class, $x_2$, is $(1.00+1.99)/2 = 1.50$, the midpoint of the third class is $(2.00+2.99)/2 = 2.50$, and so on. Then we can do the calculations in the spreadsheet shown in Figure 5.3.

| | A | B | C | D |
|---|---|---|---|---|
| 1 | **Frequency distribution** | | | |
| 2 | | | | |
| 3 | **Class** | **x** | **f** | **xf** |
| 4 | | | | |
| 5 | 0 – 0.99 | 0.5 | 1 | 0.5 |
| 6 | 1.00 – 1.99 | 1.5 | 4 | 6 |
| 7 | 2.00 – 2.99 | 2.5 | 8 | 20 |
| 8 | 3.00 – 3.99 | 3.5 | 6 | 21 |
| 9 | 4.00 – 4.99 | 4.5 | 3 | 13.5 |
| 10 | 5.00 – 5.99 | 5.5 | 1 | 5.5 |
| 11 | | | | |
| 12 | **Totals** | | 23 | 66.5 |
| 13 | **Mean** | | | 2.89 |

**Figure 5.3**   Calculation of the mean for continuous data.

The arithmetic mean usually gives a reasonable measure for location and has the advantages of:

- being easy to calculate;
- being familiar and easy to understand;
- using all the data;
- being useful in a number of other analyses;
- being objective.

It also has some weaknesses, and in particular it does not always give typical values. If four ships use a berth in eight days, the mean has the strange value of half a ship a day. Five students with exam marks of 100%, 40%, 40%, 35% and 35%, have a mean of 50% – so that four marks are below the mean while only one is above.

In general, the mean has the disadvantages of:

- only working with cardinal data;
- sometimes being misleading;
- may be some distance from most observations;
- is affected by outlying results;
- can give fractional values even with discrete data;
- gives an approximation for grouped data.

## In summary

**The most commonly used measure of location is the mean. This is calculated by $\Sigma x/n$ for ungrouped data and $\Sigma fx/\Sigma f$ for grouped data.**

### 5.2.3 Median

If you arrange a set of data in order of increasing size, the **median** is the middle value. So the median of 10, 20 and 30 is 20. In general, if there are $n$ observations the median is the value of observation $(n+1)/2$ when they are sorted into order.

## we Worked example 5.4

The times taken to inspect five units on a production line are 13, 14, 11, 17 and 11 minutes. What is the median time?

## Solution

There are 5 observations, so the median is number $(5+1)/2 = 3$ when they are sorted into order:

    11   11   **13**   14   17   median $= 13$

We have to make a small adjustment to the calculations if there is an even number of observations. Suppose Worked Example 5.4 had one more observation of 16 minutes:

    11   11   13   14   16   17

Now the middle observation is number $(6+1)/2 = 3.5$, or somewhere between 13 and 14. The usual way around this is to take the mean of these two values and give the median as $(13 + 14)/2 = 13.5$. This gives a value that did not actually occur, but is the best answer we can get.

The median is quite widely used, and has the advantages of:
- being easy to understand;
- usually giving a value that actually occurred;
- not giving fractional or impossible values;
- not being affected by outlying values;
- needing no calculation.

On the other hand it:
- can only be used with cardinal data;
- is not easy to use in other analyses.

## In summary

**When observations are sorted into order of size, the median is the middle value. This sometimes gives a more typical result than the mean.**

### 5.2.4 Mode

The mode is the value that occurs most often. If we have four values, 5, 7, 7 and 9, the value that occurs most often is 7, so this is the mode.

## we Worked example 5.5

The times taken to serve twelve customers in a shop are

  3   4   3   1   5   2   3   3   2   4   3   2 minutes

What is the mode of the times?

## Solution

Table 5.1 shows the frequency distribution of this data.

**Table 5.1**

| Class | Frequency |
|-------|-----------|
| 1 | 1 |
| 2 | 3 |
| **3** | **5** |
| 4 | 2 |
| 5 | 1 |

The most frequent value is 3, so this is the mode. This compares with a mean of 2.7 minutes and a median of 3 minutes.

Sometimes a set of data has several modes. In the following observations:

3  5  3  7  6  7  4  3  7  6  7  3  2  3  2  4  6  7  8

the most frequent values are 3 and 7, both of which appear five times. So the data has two modes, as shown in Figure 5.4.

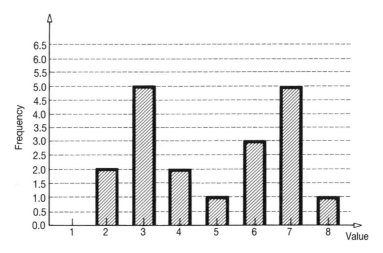

**Figure 5.4**  Frequency distribution for bimodal data.

The mode has the advantage of:

• being an actual value;

• showing the most frequent value;

• needing no calculation;

• being useful for non-numerical data.

On the other hand:

• there can be several modes;

• it cannot be used in other analyses;

• it ignores all data that are not at the mode.

Figure 5.5 shows a histogram with a typical relationship between the mean, median and mode. If the histogram is symmetrical, the three measures have the same value; but if the histogram is skewed, the measures can be a long way apart. The best one depends on both the type of data and the purpose of the information. Spreadsheets and statistical packages will do the calculations automatically, as you can see in Figure 5.6.

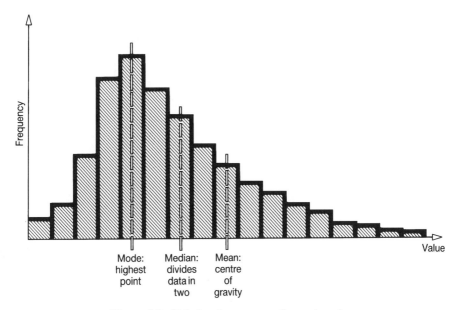

**Figure 5.5** Relating the mean, median and mode.

```
MTB   > SET C1
DATA  > 4 7 6 9 1 3 2 5 8 3 4 2
DATA  > END

MTB > MEAN C1
      MEAN  =              4.5000
MTB > MEDIAN C1
      MEDIAN  =            4.0000
MTB > MINIMUM C1
      MINIMUM  =           1.0000
MTB > MAXIMUM C1
      MAXIMUM  =           9.0000

MTB > HISTOGRAM C1

Histogram of C1    N = 12

Midpoint          Count
    1               1      *
    2               2      * *
    3               2      * *
    4               2      * *
    5               1      *
    6               1      *
    7               1      *
    8               1      *
    9               1      *
```

**Figure 5.6(a)** Minitab listing to give
the mean, median, etc. of a set of data.

|   | A | B | C | D | E | F | G | H |
|---|---|---|---|---|---|---|---|---|
| 1 | **Descriptive statistics** | | | | | | | |
| 2 | | | | | | | | |
| 3 | **Number** | **Value** | | **Description** | | | **Sorted** | |
| 4 | | | | | | | | |
| 5 | 1 | 4 | | **Number** | 12 | | 1 | |
| 6 | 2 | 7 | | **Minimum** | 1 | | 2 | Mode |
| 7 | 3 | 6 | | **Maximum** | 9 | | 2 | |
| 8 | 4 | 9 | | **Range** | 8 | | 3 | |
| 9 | 5 | 1 | | | | | 3 | |
| 10 | 6 | 3 | | | | | 4 | Median |
| 11 | 7 | 2 | | **Mean** | 4.5 | | 4 | Mean + |
| 12 | 8 | 5 | | **Median** | 4 | | 5 | |
| 13 | 9 | 8 | | **Mode** | 2 | | 6 | |
| 14 | 10 | 3 | | | | | 7 | |
| 15 | 11 | 4 | | **Class** | **Frequency** | | 8 | |
| 16 | 12 | 2 | | 1 | 1 | | 9 | |
| 17 | | | | 2 | 2 | | | |
| 18 | | | | 3 | 2 | | | |
| 19 | | | | 4 | 2 | | | |
| 20 | | | | 5 | 1 | | | |
| 21 | | | | 6 | 1 | | | |
| 22 | | | | 7 | 1 | | | |
| 23 | | | | 8 | 1 | | | |
| 24 | | | | 9 | 1 | | | |

**Figure 5.6(b)** Description of data from a spreadsheet.

## we Worked example 5.6

Two doctors and three receptionists work in a village health centre. Last year the gross annual pay earned by these five were £52 000, £48 000, £14 000, £8 000 and £8 000. How could you describe the pay? Which is the most useful measure?

## Solution

The mean pay is £26 000 a year, the median is £14 000 and the mode is £8 000. Here the mean gives a value that is not close to any observation, and the mode shows the lowest income. Although none is perfect, the median is probably the best measure.

## In summary

**The mode is the most frequently occurring value in a set of data. It can be useful, but this is probably the least popular measure of location.**

# 5.3 Measuring the spread of data

## 5.3.1 Initial measures

Averages show the location of a set of data. But they do not say how spread out the data are – and this can be very important. The mean number of customers visiting a restaurant might be 500 a week, but it is much easier to organise the restaurant for small variations – from, say, 490 on quiet weeks to 510 on busy weeks – than to deal with wide variations – between, say, 0 and 2000.

The simplest measure of spread is the **range**.

$$\text{\textbf{range}} = \text{maximum value} - \text{minimum value}$$

Unfortunately, the range can be affected by one or two extreme values. We can get around this by ignoring outlying observations. A common way of doing this uses **quartiles**, which are the values found one-quarter of the way through the data. Then:

- $Q_1$ is the first quartile, which is the value below which we find 25% of observations.

- $Q_2$ is the second quartile, which is half way through the data and is, therefore, the median.

- $Q_3$ is the third quartile, below which we find 75% of observations.

We can use these to find a narrower range $Q_3 - Q_1$ which contains 50% of observations. This has the awful name of **quartile deviation** or **semi-interquartile range**, and is calculated as:

$$\text{\textbf{quartile deviation}} = \frac{Q_3 - Q_1}{2}$$

Sometimes you will see ranges based on other calculations. The 5th percentile, for example, is the value with 5% of observations below it, while the 95th percentile has 95% of observations below it.

## In summary

**Two common measures of spread are range and quartile deviation. These are simple measures that describe the amount of dispersion in data.**

## 5.3.2 Mean absolute deviation

The quartile deviation is clearly related to the median. Other measures of spread see how far each observation is away from the mean. This is called the **deviation**.

$$\text{deviation} = \text{observation} - \text{mean value} = x_i - \bar{x}$$

Each observation has a deviation, so the mean of these deviations gives a measure of spread. Unfortunately, some of the deviations are positive and some are negative, so when we add them these positive and negative deviations cancel each other. If we have observations of 3, 4 and 8 the mean is 5, and the mean deviation is:

$$\text{mean deviation} = \frac{(3 - 5) + (4 - 5) + (8 - 5)}{3} = 0$$

So a set of dispersed data can have a mean deviation of zero. This measure is obviously unreliable, so it is never used. However, we can make a small adjustment to find the more useful **mean absolute deviation** (MAD). For this, we simply take the absolute values of deviations from the mean. In other words we ignore negative signs and add all deviations as if they are positive. Then the MAD shows the average distance of observations from the mean.

$$\text{mean absolute deviation} = \frac{\Sigma \text{ABS}(x - \bar{x})}{n}$$

where:

$x$ = the observations

$\bar{x}$ = mean value of observations

$n$ = number of observations

ABS $(x - \bar{x})$ = the absolute value of $x$, ignoring the sign

If you are doing this calculation by hand, the steps are:

● calculate the mean value;

● find the deviation of each observation from this mean;

● take the absolute values of these deviations;

● add the absolute values;

● divide this sum by the number of observations to give the MAD.

A spreadsheet can do these calculations very easily.

# we Worked example 5.7

What is the mean absolute deviation of 4, 7, 6, 10 and 8?

## Solution

The mean of the numbers is:

$$\frac{4 + 7 + 6 + 10 + 8}{5} = 7$$

Then we can find the mean absolute deviation from:

$$\text{MAD} = \frac{\text{ABS}(4–7) + \text{ABS}(7–7) + \text{ABS}(6–7) + \text{ABS}(10–7) + \text{ABS}(8–7)}{5}$$

$$= \frac{\text{ABS}(–3) + \text{ABS}(0) + \text{ABS}(–1) + \text{ABS}(3) + \text{ABS}(1)}{5}$$

$$= \frac{3 + 0 + 1 + 3 + 1}{5} = 1.6$$

This tells us that, on average, the observations are 1.6 units away from the mean.

The MAD is easy to calculate and uses all observations. It also has a clear meaning: if the MAD is 2 units, observations are an average of 2 units away from the mean. But it can be affected by a few extreme values. A more important problem is the difficulty of using it in any other statistical analysis. This rather limits its use and an alternative, the variance is much more widely used.

## In summary

**The absolute deviation is the difference between an observation and the mean. The mean absolute deviation gives a useful measure of spread.**

## 5.3.3 Variance and standard deviation

The mean absolute deviation stopped positive and negative deviations from cancelling by taking absolute values. We could also square the deviations and find the mean squared error. This gives an important measure which is called the **variance**.

$$\text{variance} = \frac{\Sigma(x - \bar{x})^2}{n}$$

Unfortunately, the units of the variance are the square of the units of the original observations. If, for example, the observations are measured in tonnes, then the variance has the meaningless units of tonnes squared. To return the units to normal, we simply take the square root of the variance. This gives the most widely used measure of spread, which is the **standard deviation**.

$$\text{standard deviation} = s = \sqrt{\frac{\Sigma(x - \bar{x})^2}{n}} = \sqrt{\text{variance}}$$

## we Worked example 5.8

Find the variance and standard deviation of 2, 3, 7, 8 and 10.

## Solution

The mean of these numbers is $(2 + 3 + 7 + 8 + 10)/5 = 6$. Then the deviations, defined as $(x - \bar{x})$, are:

$$2 - 6 = -4, \quad 3 - 6 = -3, \quad 7 - 6 = 1, \quad 8 - 6 = 2 \quad \text{and} \quad 10 - 6 = 4$$

We get the variance by squaring these deviations, adding them and dividing by the number of observations.

$$\text{variance} = \frac{\Sigma(x - \bar{x})^2}{n} = \frac{(-4)^2 + (-3)^2 + (1)^2 + (2)^2 + (4)^2}{5} = \frac{46}{5} = 9.2$$

The standard deviation is the square root of the variance:

$$\text{standard deviation} = s = \sqrt{9.2} = 3.03$$

We can extend the calculations for variance and standard deviation to grouped data using the same approach as we used for the mean. So we assume that the actual observations are at the midpoints of the classes. Then we can find the variance and standard deviation from:

$$\text{variance} = \frac{\Sigma(x - \bar{x})^2 f}{\Sigma f} = \frac{\Sigma(x - \bar{x})^2 f}{n}$$

$$\text{standard deviation} = \sqrt{\text{variance}}$$

## we Worked example 5.9

Find the variance and standard deviation of the following data.

| Class | 0-4.9 | 5-9.9 | 10-14.9 | 15-19.9 | 20-24.9 |
|-----------|-------|-------|---------|---------|---------|
| Frequency | 3 | 5 | 7 | 6 | 2 |

## Solution

Figure 5.7 shows these calculations on a spreadsheet. The mean of the 23 observations is 12.28 and this gives the deviations. The variance is 33.65, so the standard deviation is $\sqrt{33.65} = 5.8$.

| | A | B | C | D | E | F | G |
|----|----------|------|-------|--------|-----------|-----------|-------------|
| 1 | **Variance** | | | | | | |
| 2 | | | | | | | |
| 3 | **Class** | **x** | **f** | **f*x** | **deviation** | **dev.sqrd.** | **f* dev.sqrd** |
| 4 | | | | | | | |
| 5 | **0–4.9** | 2.5 | 3 | 7.5 | −9.78 | 95.70 | 287.10 |
| 6 | **5–9.9** | 7.5 | 5 | 37.5 | −4.78 | 22.87 | 114.37 |
| 7 | **10–14.9** | 12.5 | 7 | 87.5 | 0.22 | 0.05 | 0.33 |
| 8 | **15–19.9** | 17.5 | 6 | 105 | 5.22 | 27.22 | 163.33 |
| 9 | **20–24.9** | 22.5 | 2 | 45 | 10.22 | 104.40 | 208.79 |
| 10 | | | | | | | |
| 11 | **Sums** | | 23.00 | 282.50 | | | 773.91 |
| 12 | **Means** | | | 12.28 | | | 33.65 |

**Figure 5.7** Calculation of variance and standard deviation for grouped data.

The variance and standard deviation do not have direct meanings like the mean absolute deviation. We can only say that a large variance shows more spread than a smaller one – so data with a variance of 42.5 is more spread out than equivalent data with a variance of 22.5.

One important point about variances is that they can sometimes be added. Provided that two sets of observations are completely unrelated, the variance of the sum of data is equal to the sum of the variances of each set. If, for example, the daily demand for an item has a variance of 4, while the daily demand for a second item has a variance of 5, the variance of total demand for both items is $4 + 5 = 9$. You can never add standard deviations in this way.

> ## we Worked example 5.10

Alabanic Airlines knows that the mean weight and standard deviation of its passengers are 72 kg and 6 kg respectively. If 200 passengers board a plane, how could you describe the total weight?

## Solution

You can find the total weight of 200 passengers by multiplying the mean weight of each passenger by 200:

mean total weight $= 200 \times 72 = 14\,400$ kg

Similarly, you can find the total variance by multiplying the variance of each passenger's weight by 200.

total variance $= 200 \times 6^2 = 7200$ kg

The standard deviation in total weight is $\sqrt{7200} = 84.85$ kg.

### In summary

**The most widely used measure of data spread is the variance, which is the mean squared deviation. The square root of this is the standard deviation.**

# 5.4 Using indices to describe changes

## 5.4.1 Introduction

In business, many values change over time such as the price paid for material, the number of people employed, the annual profit and the number of customers. We can plot graphs to show these trends, but this does *not* give a **measure** for the changes. One way of measuring changes uses an **index** or **index number**.

An index is a number that compares the value of a variable at any time with its value at another fixed time, called the **base period**. Then we can define:

$$\textbf{index for period} = \frac{\text{value in period}}{\text{value in base period}}$$

## we Worked example 5.11

A police force recorded 127 crimes of a particular type in one year, 142 crimes in the second year, and 116 crimes in the third year. Use the first year as a base period to give the indices for crimes in the three years.

## Solution

These calculations are shown in Table 5.2.

**Table 5.2**

| Year | Value | Calculation | Index |
|------|-------|-------------|-------|
| 1 | 127 | $\dfrac{127}{127}$ | 1.00 |
| 2 | 142 | $\dfrac{142}{127}$ | 1.12 |
| 3 | 116 | $\dfrac{116}{127}$ | 0.91 |

The index in the base period is always 1.00. The index of 1.12 in the second year shows that crime is 12% higher than the **base value** (which is the value in the base period), while the index of 0.91 in the third period tells us that crime is 9% lower than the base value.

An important use of indices shows how the price of a product changes over time. For convenience, these indices are usually multiplied by 100, so we get:

$$\text{Price index in period } N = \frac{\text{price in period } N}{\text{base price}} \times 100$$

## we Worked example 5.12

If the price of an item is £20 in January, £22 in February, £25 in March, and £26 in April, what is the price index in each month using January as the base month?

## Solution

Taking January as the base month, the base price is £20. Then the price indices for each month are:

January: $\dfrac{20}{20} \times 100 = 100$ (as expected in the base period)

February: $\dfrac{22}{20} \times 100 = 110$

March: $\dfrac{25}{20} \times 100 = 125$

April: $\dfrac{26}{20} \times 100 = 130$

In Worked Example 5.12, the index of 110 in February shows that the price has risen by 10% over the base level, while an index of 125 in March shows a rise of 25% over the base level, and so on. Changes in indices between periods are called **percentage point** changes. So between February and March the index shows an increase of $125 - 110 = 15$ percentage points (compared with a percentage increase of $15/110 \times 100 = 13.6\%$). Remember that percentage point changes refer back to the base price and not the current price.

## we Worked example 5.13

The following table shows the monthly price index for an item.

| Month | 1 | 2 | 3 | 4 | 5 | 6 | 7 | 8 | 9 |
|---|---|---|---|---|---|---|---|---|---|
| Index | 121 | 112 | 98 | 81 | 63 | 57 | 89 | 109 | 131 |

(a) If the price in month 3 is £240 what is the price in month 8?

(b) What is the percentage point rise between months 7 and 8?

## Solution

(a) In month 3 the price is £240 and the price index is 98, so:

$$\text{base price} = \dfrac{240}{98} \times 100 = 244.90$$

Then in month 8 the price index is 109 so the price is:

$$\text{price} = 244.90 \times \frac{109}{100} = 266.94$$

(b) The indices in periods 7 and 8 are 89 and 109 respectively. The percentage point rise is the difference between these, which is 20 percentage points.

## In summary

**In business, many values change over time. We can use an index to measure these changes. An index gives the ratio of the current value over a base value.**

## 5.4.2 Changing the base period

Rather than keep the same base period for a long time, it is usual to update it periodically. There are two reasons for this:

- **Changing circumstances**. It makes sense to reset the index whenever there are significant changes in circumstances. So a manufacturer who uses an index for production might return the base value to 100 whenever the product range changes.

- **The index get too big**. If an index gets to, say 10 000, a 10% increase will move it to 11 000 and this seems a much bigger step than from 100 to 110.

Suppose we have an old index that uses a base value in period $N_1$, and replace this by a new index that uses a base value in period $N_2$. How can we update the previous indices? The answer comes from:

$$\textbf{new index} = \text{old index} \times \frac{\text{value in period } N_1}{\text{value in period } N_2}$$

As the values in both periods $N_1$ and $N_2$ are fixed, we can find the new index by multiplying the old index by a fixed amount.

## we Worked example 5.14

The number of units made each year in a factory is described by the following indices.

| Year | 1 | 2 | 3 | 4 | 5 | 6 | 7 | 8 |
|------|-----|-----|-----|-----|-----|-----|-----|-----|
| Index 1 | 100 | 138 | 162 | 196 | 220 | | | |
| Index 2 | | | | | 100 | 125 | 140 | 165 |

(a) If the factory had not changed to index 2, what values would index 1 have had in years 6 to 8?

(b) What values would index 2 have had in years 2 to 4?

(c) If the factory made 4860 units in year 3, how many did it make in year 7?

## Solution

(a) We can find index 1 by multiplying index 2 by a constant amount. You can see from year 5 that this constant is 220/100. So the values for index 1 are:

year 6: $125 \times \dfrac{220}{100} = 275$

year 7: $140 \times \dfrac{220}{100} = 308$

year 8: $165 \times \dfrac{220}{100} = 363$

(b) Again, we can find index 2 by multiplying index 1 by a constant amount. You can see from year 5 that this constant is now 100/220. So the values for index 2 are:

year 4: $196 \times \dfrac{100}{220} = 89.09$

year 3: $162 \times \dfrac{100}{220} = 73.64$

year 2: $138 \times \dfrac{100}{220} = 62.73$

(c) The factory made 4860 units in year 3, when index 1 had a value of 162. So in year 7, when index 1 had a value of 308, the production is:

$$4860 \times \frac{308}{162} = 9240 \text{ units}$$

We can check this result using index 2. The factory made 4860 units in year 3, when index 2 had a value of 73.64. So in year 7, when index 2 had a value of 140, the production is:

$$4860 \times \frac{140}{74.64} = 9240 \text{ units}$$

## In summary

**Base periods for index numbers should be changed periodically. We can calculate a new index by multiplying the old index by a constant.**

### 5.4.3 Indices for more than one variable

Often, we do not want to see how a single variable changes over time, but how a combination of different variables changes. The retail price index shows how the total price of a range of goods varies over time; a diversified company wants to see how the performance of all of its divisions changes; a local authority wants to plot changes in the number of people it employs in different departments, and so on. Indices which measure changes in a number of variables are called **aggregate indices**. A reasonable aggregate index must take into account two factors:

- the way each individual variable changes;
- the weights given to each variable.

There are several ways of combining these into a single index. If you want to see how the amount that a family pays for food is changing, you can simply record the price of each week's shopping basket and use an index of the total cost. The total cost depends on both the price of each item and the number of items bought, so we have an aggregate index:

$$\text{aggregate price index} = \frac{\text{cost of week's shopping basket}}{\text{cost of shopping basket in base period}}$$

Unfortunately, this index has a major weakness. If some prices change, the amount the family buys will also change. If the price of cake goes up, then the family may buy less cake and more biscuits. So the index should take these changes into account. The easiest ways of doing this use base period weighting or current period weighting.

**Base weighted index**

A base weighted index assumes that the amounts bought do not change from the base period:

$$\text{base weighted index} = \frac{\text{cost of base period quantities at current prices}}{\text{cost of base period quantities at base period prices}}$$

This index has the advantage of giving a direct comparison of costs, but it assumes that the amounts bought do not change. In practice, purchases are affected by prices, and a product whose price rises quickly will be replaced by one with a lower price. As a result, base weighted indices tend to give values that are too high.

**Current period weighted index**

A current period weighted index assumes that the current shopping basket was used in the base period.

$$\text{current period weighted index} = \frac{\text{cost of current quantities at current prices}}{\text{cost of current quantities at base period prices}}$$

This index measures changes in the costs of current purchases. But it changes the mix of purchases each period, so the index does not give a direct comparison of prices over time. By substituting products that are cheaper than they were in the base period, the index tends to be too low.

## we Worked example 5.15

A company buys four products with the following characteristics.

**Table 5.3**

| Item | Number of units bought | | Price paid per unit | |
|------|------|------|------|------|
|      | Year 1 | Year 2 | Year 1 | Year 2 |
| A | 20 | 24 | 10 | 11 |
| B | 55 | 51 | 23 | 25 |
| C | 63 | 84 | 17 | 17 |
| D | 28 | 34 | 19 | 20 |

(a) Find the price indices for each product in year 2 using year 1 as the base year.

(b) Calculate a base weighted index for the products.

(c) Calculate a current period weighted index.

## Solution

(a) Simple price indices do not take into account usage of a product, so we have:

product A: $\dfrac{11}{10} \times 100 = 110.0$

product B: $\dfrac{25}{23} \times 100 = 108.7$

product C: $\dfrac{17}{17} \times 100 = 100.0$

product D: $\dfrac{20}{19} \times 100 = 105.3$

(b) A base weighted index shows the price that would be paid later for the basket of items bought in the base period. So using the amounts bought in year 1 gives:

$$\text{base weighted index} = \frac{(11 \times 20) + (25 \times 55) + (17 \times 63) + (20 \times 28)}{(10 \times 20) + (23 \times 55) + (17 \times 63) + (19 \times 28)}$$

$$= \frac{3226}{3068} = 105.15$$

(c) A current period weighted index shows how the price of the basket of items bought in a later period has changed since the base period. So using the amounts bought in year 2 gives:

$$\text{current period weighted index} = \frac{(11 \times 24) + (25 \times 51) + (17 \times 84) + (20 \times 34)}{(10 \times 24) + (23 \times 51) + (17 \times 84) + (19 \times 34)}$$

$$= \frac{3647}{3487} = 104.59$$

## In summary

An aggregate index shows changes in a combination of variables. It can show changes in both price and quantities bought. The two most common aggregate indices use base weighting and current period weighting.

## Chapter review

This chapter showed how data can be described numerically. It concentrated on measures for location, spread and changes. In particular it:

- discussed the need for numerical descriptions of data;
- described the mean, median and mode for describing the location of data;
- looked at the range and quartile deviation to measure data spread;
- calculated mean absolute deviation, variance and standard deviation;
- used simple and aggregate indices to describe changes to data.

## Problems

5.1 The number of visitors received by a hospital patient in five days were 4, 2, 1, 7 and 1. What are the mean, median and mode of numbers visiting?

5.2 Find the mean, median and mode of the following numbers:

24  26  23  24  23  24  27  26  28  25  21  22  25  23  26  29  27  24  25  24  24  25

5.3 Find the mean of the frequency distribution shown in the following table.

| Class | 1.00–2.99 | 3.00–4.99 | 5.00–6.99 | 7.00–8.99 | 9.00–10.99 | 11.00–12.99 |
|---|---|---|---|---|---|---|
| Frequency | 2 | 6 | 15 | 22 | 12 | 4 |

5.4 What are the mean absolute deviation, variance and standard deviation of:

27, 32, 34, 28, 35, 30.

5.5 How could you measure the spread of the data described in Problems 5.1 and 5.2?

5.6 A hotel is concerned about the number of people who book rooms by telephone but do not actually turn up. Over the past few weeks it has kept records of the number of people who do this, as shown overleaf. How can you summarise these data?

| Day | 1 | 2 | 3 | 4 | 5 | 6 | 7 | 8 | 9 | 10 | 11 | 12 | 13 | 14 | 15 |
|-----|---|---|---|---|---|---|---|---|---|----|----|----|----|----|----|
| No-shows | 4 | 5 | 2 | 3 | 3 | 2 | 1 | 4 | 7 | 2 | 0 | 3 | 1 | 4 | 5 |

| Day | 16 | 17 | 18 | 19 | 20 | 21 | 22 | 23 | 24 | 25 | 26 | 27 | 28 | 29 | 30 |
|-----|----|----|----|----|----|----|----|----|----|----|----|----|----|----|----|
| No-shows | 2 | 6 | 2 | 3 | 3 | 4 | 2 | 5 | 5 | 2 | 4 | 3 | 3 | 1 | 4 |

| Day | 31 | 32 | 33 | 34 | 35 | 36 | 37 | 38 | 39 | 40 | 41 | 42 | 43 | 44 | 45 |
|-----|----|----|----|----|----|----|----|----|----|----|----|----|----|----|----|
| No-shows | 5 | 3 | 6 | 4 | 3 | 1 | 4 | 5 | 6 | 3 | 3 | 2 | 4 | 3 | 4 |

5.7    Figure 5.8 shows a spreadsheet describing some data. What do these results show?

| | A | B | C | D | E | F | G |
|---|---|---|---|---|---|---|---|
| 1 | **Data description** | | | | | | |
| 2 | | | | | | | |
| 3 | **Data** | | | | | **Description** | |
| 4 | | | | | | | |
| 5 | 15 | 38 | 26 | 42 | | Mean | 34.85 |
| 6 | 23 | 40 | 32 | 51 | | Median | 38 |
| 7 | 41 | 37 | 40 | 47 | | Mode | 40 |
| 8 | 16 | 52 | 50 | 51 | | Standard deviation | 12.73 |
| 9 | 32 | 20 | 18 | 30 | | Variance | 162.05 |
| 10 | 29 | 33 | 24 | 20 | | Kurtosis | -1.06 |
| 11 | 10 | 14 | 27 | 39 | | Skewness | -0.24 |
| 12 | 52 | 57 | 45 | 48 | | Range | 47 |
| 13 | 17 | 48 | 40 | 38 | | Minimum | 10 |
| 14 | 23 | 39 | 46 | 44 | | Maximum | 57 |
| 15 | | | | | | Sum | 1394 |
| 16 | | | | | | Count | 40 |

**Figure 5.8**    Description of data in a spreadsheet.

5.8    The number of fishing boats working from a harbour over the past ten years is as follows:

325   321   316   294   263   241   197   148   102   70

Describe these changes by indices based on the first and last years' observations.

5.9    An insurance company uses an index to describe the number of agents working for it. This index was revised five years ago, and its values over the past ten years are shown in Table 5.4.

**Table 5.4**

| Year | 1 | 2 | 3 | 4 | 5 | 6 | 7 | 8 | 9 | 10 |
|------|---|---|---|---|---|---|---|---|---|----|
| Index 1 | 106 | 129 | 154 | 173 | 195 | 231 | | | | |
| Index 2 | | | | | | 100 | 113 | 126 | 153 | 172 |

(a) If the company had not changed to index 2, what values would index 1 have in years 7 to 10?

(b) What values would index 2 have in years 1 to 5?

(c) If the company had 645 agents in year 4, how many did it have in each other year?

5.11 A company buys four products whose characteristics are shown in Table 5.5.

**Table 5.5**

| Products | Number of units bought | | Price paid per unit | |
|---|---|---|---|---|
| | Year 1 | Year 2 | Year 1 | Year 2 |
| A | 121 | 141 | 9 | 10 |
| B | 149 | 163 | 21 | 23 |
| C | 173 | 182 | 26 | 27 |
| D | 194 | 103 | 31 | 33 |

Calculate a base weighted index and a current period weighted index for the products.

# CS  Case study – Consumer Advice Office

Mary Smith has been working in the local authority as a Consumer Advice Officer for the past fourteen months. Her job is to advise people who complain about local traders. She listens to the complaints, assesses the problem and then takes any follow-up action. She can often deal with clients quickly, but sometimes there is a lot of follow-up including legal work and appearances in court.

The local authority is always looking for ways to reduce costs and improve its service, so it is important for Mary to show that she is doing a good job. She is particularly keen to show that she is responding to pressures for greater efficiency by dealing with more clients.

Mary has records of the number of clients she dealt with during her first eight weeks at work, and during the same eight weeks this year:

● Number of customers dealt with each working day in the first eight weeks:

  6  18  22   9  10  14  22  15  28   9  30  26  17   9  11  25  31  17  25  30
 32  17  27  34  15   9   7  10  28  10  31  12  16  26  21  37  25   7  36  29

● Number of customers dealt with each working day in the last eight weeks:

 30  26  40  19  26  31  28  41  18  27  29  30  33  43  19  20  44  37  29  22
 41  39  15   9  22  26  30  35  38  26  19  25  33  39  31  30  20  34  43  45

During the past year she estimates that she has worked an extra two hours a week as unpaid overtime. Her wages increased by 3% after allowing for inflation. What she needs is a way of presenting these figures to her employers in a form that they will understand. How can she do this?

part

# 3

# *Solving business problems*

We have looked in Parts One and Two at the background to quantitative methods and at data collection and description. Part Three now shows how to solve some specific types of business problem.

There are four chapters in Part Three. The first chapter describes some financial calculations. The next three chapters look at forecasting and some other types of problem.

# 6 Calculations with money

## Contents

## Chapter outline

This chapter describes some financial analyses that are widely used in business. After reading the chapter you should be able to:

- see why organisations need measures of financial performance
- calculate break-even points
- understand some reasons for economies of scale
- do calculations with interest rates
- calculate present values and internal rates of return
- depreciate the value of assets
- calculate payments for sinking funds and mortgages

# 6.1 Financial measures

All organisations have certain financial objectives. Companies want to make a profit, charities want to collect as much money as possible, hospitals want to treat all patients with acceptable costs, and schools want to give good education with limited resources. So organisations need some way of measuring financial performance. There are several measures available, but perhaps the most widely used is **return on assets**.

$$\textbf{return on assets} = \frac{\text{income}}{\text{total assets}}$$

From a financial view, the return on assets should be as high as possible, so 15% is better than 10%. Another widely used measure is the price/earnings ratio:

$$\textbf{price/earnings ratio} = \frac{\text{share price}}{\text{earnings per share}}$$

There are many other measures of performance, some of which we describe in the rest of the chapter.

## In summary

**All organisations need measures for their financial performance. There are many possible such measures.**

# 6.2 Break-even point

## 6.2.1 Making enough for a profit

Organisations usually try to make a profit.

$$\textbf{profit} = \text{revenue} - \text{costs}$$

Revenue depends on the number of units sold. So a useful calculation checks that a company will sell enough units to make a profit. We describe this break-even analysis in terms of a company making a product, but this is just for convenience and we can use the results in many different circumstances.

When a company sells a product for a fixed price, it generates an income or revenue of:

revenue = price per unit $\times$ number of units sold = $PN$

where:

$P$ = price charged for each unit
$N$ = number of units sold

The cost of making the product is a little more awkward as some costs are fixed regardless of the number of units made, while others vary with the output. If, for example, a company hires a machine to make a product, the cost of hiring may be fixed regardless of the number of units made, while the cost of raw materials will vary. You can see this effect with a car. Some of the costs are fixed regardless of the distance travelled (repayment of the purchase loan, road tax, insurance, etc.) while other costs vary with the mileage (petrol, oil, tyres, depreciation, etc.). So we can describe:

total cost = fixed cost + variable cost

= fixed cost + (cost per unit $\times$ number of units made)

= $F + CN$

where:

$F$ = fixed cost

$C$ = cost per unit

$N$ = number of units made

Now we can compare the cost of making $N$ units of a product with the income from selling them. In particular, we can find a **break-even point**, which is the number of units that must be sold to cover all costs and make a profit.

Suppose a company has to spend £200 000 on research, development, equipment and other preparations before it can start making a new product. During normal production, each unit of the new product costs £20 to make and sells for £30. The company will only start to make a profit when it has recovered the original £200 000. As each unit contributes £30 − £20 = £10, the company must sell

$$\frac{200\,000}{10} = 20\,000 \text{ units}$$

to recover the original investment. This is the break-even point.

At the break-even point the revenue covers all costs, so:

revenue = total cost

But we have already found these, so, assuming that all production is sold:

$PN = F + CN$

or:

$N(P - C) = F$

Then:

$$\text{break-even point} = N = \frac{F}{P - C}$$

You can see that both the revenue and total cost rise linearly with the number of units, so we can plot the relationships as the graph in Figure 6.1.

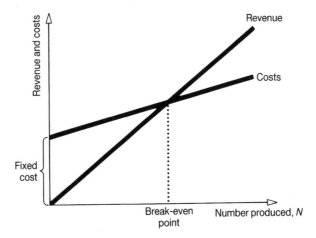

**Figure 6.1** Illustrating the break-even point.

## we Worked example 6.1

Belmonte Industrial has fixed costs for buildings, machines and employees of £12 000 a week. Raw material and other variable costs are £50 a unit.

(a) What is the break-even point if the selling price is £130 a unit? What is the profit if sales remain are 200 units a week?

(b) What is the break-even point if the selling price is reduced to £80 a unit? What is the profit if sales remain at 200 units a week?

(c) What is the profit if the lower price increases sales to 450 units a week?

## Solution

(a) We know that:

$F = £12\,000$ a week $=$ fixed cost each week

$C = £50$ a unit $\qquad =$ variable cost per unit

$P = £130$ a unit $\qquad =$ price per unit

The break-even point is:

$$N = \frac{F}{P-C} = \frac{12\,000}{130-50} = 150 \text{ units}$$

Actual sales are more than the break-even point, so the product makes a profit of:

$$\text{profit} = \text{revenue} - (\text{fixed cost} + \text{variable cost})$$

$$= PN - (F + CN)$$

$$= (130 \times 200) - \{12\,000 + (50 \times 200)\} = \pounds4000 \text{ a week}$$

(b) With a selling price $P$ of £80 the break-even point is:

$$N = \frac{F}{P - C} = \frac{12\,000}{80 - 50} = 400 \text{ units}$$

Actual sales are less than this, so the income will not cover the costs:

$$\text{profit} = \text{revenue} - (\text{fixed cost} + \text{variable cost})$$

$$= PN - (F + CN)$$

$$= (50 \times 200) - \{12\,000 + (80 \times 200)\} = -\pounds6000 \text{ a week}$$

(c) With the selling price set at £80 a unit, the break-even point is 400 units. If sales increase to 450 units a week the company makes a profit of:

$$\text{profit} = \text{revenue} - (\text{fixed cost} + \text{variable cost})$$

$$= PN - (F + CN)$$

$$= (80 \times 450) - \{12\,000 + (50 \times 450)\} = \pounds1500 \text{ a week}$$

A common difficulty in calculating break-even points is assigning a reasonable proportion of overheads to the fixed cost of each product. The proportion assigned depends on the accounting conventions used within the organisation. The problem is made worse if the product mix is constantly changing: then accounting practices will allocate a changing amount of overheads to each product. In other words, the costs of making a particular product can change, even though there has been no change in the product itself or the way it is made.

## In summary

**Costs can be classified as either fixed or variable. Revenue must cover both of these before giving a profit – and this occurs at the break-even point.**

### 6.2.2 Economies of scale

Break-even analysis shows one reason why organisations can get **economies of scale**. These economies allow the average cost per unit to fall as the number of units sold increases. We know that:

$$\text{total cost} = \text{fixed cost} + \text{variable cost} = F + CN$$

Dividing this total cost by the number of units, $N$, gives the average cost per unit:

$$\text{average cost per unit} = \frac{F + CN}{N} = \frac{F}{N} + C$$

As $N$ increases, so the average cost per unit falls, because the amount of the fixed cost recovered by each unit is reduced.

## we Worked example 6.2

A restaurant serves 200 meals a day at an average price of £20. The variable cost of each meal is £10, and the fixed costs of running the restaurant are £1750 a day.

(a) What is the break-even point? How much profit does the restaurant make?

(b) What is the average cost of a meal?

(c) If the number of meals rises to 250 a day, how much does the average cost of a meal fall?

## Solution

(a) The break-even point is:

$$N = \frac{F}{P - C} = \frac{1750}{20 - 10} = 175$$

Actual sales are above this so the restaurant makes a profit of:

$$\text{profit} = \text{revenue} - (\text{fixed cost} + \text{variable cost})$$

$$= PN - (F + CN)$$

$$= (20 \times 200) - \{1750 + (10 \times 200)\} = £250 \text{ a day}$$

(b) The average total cost of a meal is:

$$\text{average cost} = \frac{F}{N} + C = \frac{1750}{200} + 10 = £18.75 \text{ a meal}$$

(c) Serving 250 meals a day gives an average cost of:

$$\frac{F}{N} + C = \frac{1750}{250} + 10 = £17 \text{ a meal}$$

Spreading the fixed costs over more units is only one reason for economies of scale. There are often economies of scale even when fixed costs are ignored. These also happen because people become more familiar with the operations and work faster; machine operators become more practised; more specialised equipment can be used; distribution costs are reduced; raw materials can be bought with a discount; disruptions are eliminated, and so on.

People often think that there are always economies of scale, and operations should be as big as possible – but there can also be **diseconomies of scale**. These happen when the savings are more than offset by increased bureaucracy, difficulties of communication, more complex management hierarchies, increased costs of supervision, perceived reduction in the importance of individuals, and so on. This usually gives economies of scale up to an optimal size, and then diseconomies of scale as inefficiencies set in, causing the average cost per unit to rise, as shown in Figure 6.2.

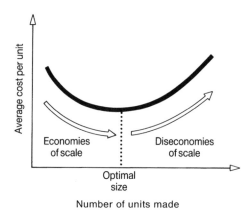

**Figure 6.2**   Economies and diseconomies of scale.

## In summary

**Economies of scale can often reduce the average unit costs, but beyond a certain point there may be diseconomies of scale.**

# 6.3 Value of money over time

## 6.3.1 Interest rates

If you put some money into a bank account, it earns interest. Interest rates are usually quoted as annual percentages, like 10%. But if you leave £1000 in an account earning interest at 10% a year, it will earn $1000 \times \frac{10}{100} = £100$ at the end of the year. It is easier to calculate interest using a decimal fraction, $i$, rather than a percentage. So an interest rate of 10% has $i = 0.1$, an interest rate of 15% has $i = 0.15$, and so on. Then if you put an amount of money, $A_0$, into a bank account and it earns interest at an annual rate, $i$, at the end of the year there will be an amount $A_1$ in the account, where:

$$A_1 = A_0 \times (1 + i)$$

If you leave the amount $A_1$ for a second year, it will earn interest on both the initial amount and the interest earned in the first year. So:

$$A_2 = A_1 \times (1 + i)$$

If we substitute the value of $A_1$:

$$A_2 = \{A_0 \times (1 + i)\} \times (1 + i)$$
$$= A_0 (1 + i)^2$$

The amount of money increases in this compound way, so that at the end of three years there will be $A_0 \times (1 + i)^3$ in the account, and at any time $n$ years in the future the account will contain $A_n$.

> with **compound interest:** $A_n = A_0 \times (1 + i)^n$

So, The £1000 will grow to:

- at the end of year 1: $1000 \times (1 + 0.1)^1 = 1000 \times 1.1 = £1100$;
- at the end of year 2: $1000 \times (1 + 0.1)^2 = 1000 \times 2.1 = £1210$;
- at the end of year 3: $1000 \times (1 + 0.1)^3 = 1000 \times 1.331 = £1331$;

and so on.

Figure 6.3 shows the value of $A_n$ for an investment of £1 and interest rates between 3% and 25%.

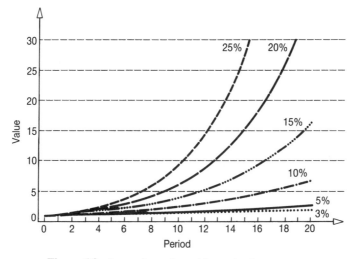

**Figure 6.3**   Increasing value with varying interest rates.

## **We** Worked example 6.3

Jane has placed £1000 in a bank account earning 5% interest a year. How much will she have in the account at the end of five years? How much will she have at the end of twenty years?

## Solution

We know that $A_0 = £1000$ and $i = 0.05$. We can calculate the amount in the account from:

$$A_n = A_0 \times (1 + i)^n$$

At the end of five years there will be:

$$A_5 = 1000 \times (1 + 0.05)^5 = 1000 \times 1.2763 = £1276$$

At the end of twenty years there will be:

$$A_{20} = 1000 \times (1 + 0.05)^{20} = 1000 \times 2.6533 = £2653$$

People lending money have to quote an APR, or **annual percentage rate** of interest, which gives the true cost of borrowing – but you have to be careful with the calculations. If you borrow £100 with interest of 2% a month, you might think that this is the same as $2 \times 12 = 24\%$ a year, but:

● borrowing £100 at 24% APR would raise the debt to

$$100 \times (1 + 0.24) = £124$$

at the end of the year, while

● borrowing £100 at 2% a month raises the debt at the end of 12 months to:

$$A_n = A_0 \times (1 + i)^n = 100 \times (1 + 0.02)^{12} = £126.82$$

So an interest rate of 2% a month is equivalent to an APR of 26.82%.

## In summary

**The value of money varies over time. In particular, an amount at present can earn interest to give a larger amount in the future.**

### 6.3.2 Present value of money

In the previous section we saw that an amount of money, $A_0$, invested now will have a value of $A_0 \times (1 + i)^n$ at a time $n$ periods in the future. We can turn this the other way around and say that an amount, $A_n$, $n$ periods in the future has a present value of $A_0$. This is expressed as:

$$A_0 = \frac{A_n}{(1+i)^n} = A_n \times (1+i)^{-n}$$

Finding the present value of an amount in the future is called **discounting to present value**. We can use this to compare amounts of money at different times.

## we Worked example 6.4

A company can do one of two projects. The profits from these are phased over many years, but they can be summarised as:

● project 1 gives £300 000 in five years' time

● project 2 gives £500 000 in ten years' time

Which project should the company do if it uses a discounting rate of 20% a year for future revenues?

## Solution

The company has to compare amounts of money earned at different times. The best way of doing this is to calculate the present value of each amount.

*Project 1:*  $A_0 = A_n \times (1+i)^{-n} = 300\,000 \times (1+0.2)^{-5}$

$= £120\,563$

*Project 2:*  $A_0 = A_n \times (1+i)^{-n} = 500\,000 \times (1+0.2)^{-10}$

$= £80\,753$

Now we have the present values of both projects. The better project is the one with the higher value, which is project 1.

Discounting to present values is particularly useful with large projects that have payments and incomes spread over varying periods. Then we can discount all amounts to their present value, and subtract the present value of all costs from the present value of all revenues to give a **net present value** (NPV).

> **net present** = sum of discounted − sum of discounted
> **value**           revenues             costs

If the NPV is negative a project will make a loss and it should not go ahead. If alternative projects have positive NPVs, the best is the one with the highest NPV.

## we Worked example 6.5

A company is considering three alternative projects with initial costs and projected revenues (each in thousands of pounds) for the next five years as shown in Table 6.1. The company can start only one project: use a discounting rate of 10% to find the best.

**Table 6.1**

| | Initial cost | Net revenue generated in each year | | | | |
|---|---|---|---|---|---|---|
| Project | | 1 | 2 | 3 | 4 | 5 |
| A | 1000 | 500 | 400 | 300 | 200 | 100 |
| B | 1000 | 200 | 200 | 300 | 400 | 400 |
| C | 500 | 50 | 200 | 200 | 100 | 50 |

## Solution

To compare these projects we want to find the net present value of each.

*Project A:*

500 in year 1 has a present value of $500/1.1 = 454.545$

400 in year 2 has a present value of $400/1.1^2 = 330.579$

300 in year 3 has a present value of $300/1.1^3 = 225.394$    and so on.

Adding these discounted revenues, and subtracting the discounted costs gives the net present value of each project. We can easily do these calculations using a spreadsheet, as shown in Figure 6.4

| | A | B | C | D | E | F | G | H |
|---|---|---|---|---|---|---|---|---|
| 1 | **Net present values** | | | | | | | |
| 2 | | | | | | | | |
| 3 | | | **Project A** | | **Project B** | | **Project C** | |
| 4 | **Year** | **Discount factor** | **Revenue** | **Present value** | **Revenue** | **Present value** | **Revenue** | **Present value** |
| 5 | 1 | 1.1 | 500 | 454.55 | 200 | 181.82 | 50 | 45.45 |
| 6 | 2 | 1.21 | 400 | 330.58 | 200 | 165.29 | 200 | 165.29 |
| 7 | 3 | 1.331 | 300 | 225.39 | 300 | 225.39 | 200 | 150.26 |
| 8 | 4 | 1.4641 | 200 | 136.60 | 400 | 273.21 | 100 | 68.30 |
| 9 | 5 | 1.61051 | 100 | 62.09 | 400 | 248.37 | 50 | 31.05 |
| 10 | **Totals** | | **1500** | **1209.21** | **1500** | **1094.08** | **600** | **460.35** |
| 11 | | | | | | | | |
| 12 | **Present values** | | | | | | | |
| 13 | | **Revenues** | | 1209.21 | | 1094.08 | | 460.35 |
| 14 | | **Costs** | | 1000 | | 1000 | | 500 |
| 15 | | | | | | | | |
| 16 | **Net present values** | | | **209.21** | | **94.08** | | **−39.65** |

**Figure 6.4**   Net present values for Worked Example 6.5.

You can see that Project A has the highest NPV and is the most attractive. Project C has a negative NPV – showing a loss – and should be avoided.

## In summary

**We can compare revenues and payments made at different times by discounting amounts to their present values. The overall benefit is measured by the net present value.**

## 6.3.3  Internal rate of return

In Worked Example 6.5 we compared three projects by using a fixed discounting rate to give three net present values. It is often difficult to find a reasonable discounting rate that takes into account interest rates, inflation, taxes, opportunity costs, exchange rates, risk and everything else. An alternative is to find the discounting rates that lead to a specified net present value. In other words, we use the same net present value for each project and then find three different discounting rates. Usually we use the discounting rate that gives a net present value of zero, and this is called the **internal rate of return** (IRR). The best project has the highest internal rate of return.

> The **internal rate of return** is the discounting rate that gives a net present value of zero.

One small difficulty with internal rates of return is that there is no simple formula for calculating them. In practice, of course, this makes no difference as computers will always do the arithmetic.

## we Worked example 6.6

What are the internal rates of return for the following two cash flows?

| Year | 0 | 1 | 2 | 3 | 4 | 5 | 6 | 7 | 8 |
|---|---|---|---|---|---|---|---|---|---|
| Net cash flow 1 | −2000 | −5000 | −200 | 800 | 1800 | 1600 | 1500 | 200 | 100 |
| Net cash flow 2 | −5000 | −4000 | −2000 | 0 | 2000 | 3000 | 3000 | 2000 | 2000 |

## Solution

The easiest way to find the internal rate of return is to use a spreadsheet. Figure 6.5 shows a typical printout. This calculates the internal rates of return as −4.376% (which gives a loss) and 1.694% (which gives a small profit).

| | A | B | C | D | E | F |
|---|---|---|---|---|---|---|
| 1 | **Internal rate of return** | | | | | |
| 2 | | | | | | |
| 3 | | **Project 1** | | | **Project 2** | |
| 4 | **Year** | **Net cash flow** | **Discounted value** | | **Net cash flow** | **Discounted value** |
| 5 | 0 | −2000 | −2000.000 | | −5000 | −5000.000 |
| 6 | 1 | −5000 | −5228.822 | | −4000 | −3933.364 |
| 7 | 2 | −200 | −218.725 | | −2000 | −1933.919 |
| 8 | 3 | 800 | 914.938 | | 0 | 0.000 |
| 9 | 4 | 1800 | 2152.821 | | 2000 | 1870.022 |
| 10 | 5 | 1600 | 2001.195 | | 3000 | 2758.304 |
| 11 | 6 | 1500 | 1961.980 | | 3000 | 2712.354 |
| 12 | 7 | 200 | 273.569 | | 2000 | 1778.113 |
| 13 | 8 | 100 | 143.044 | | 2000 | 1748.491 |
| 14 | **IRR** | **−4.376%** | | | **1.694%** | |
| 15 | **NPV** | | **0.000** | | | **0.000** |

**Figure 6.5**   Internal rates of return for Worked Example 6.6.

## In summary

**The internal rate of return is the discounting rate which gives a net present value of zero. The best investment is the one with the highest internal rate of return.**

# 6.4 Depreciation

When people buy a car, they expect to drive it for a few years and then replace it. This is because maintenance and repair costs rise and there are more breakdowns and, as the car gets older new cars are more efficient and comfortable, and so on. So the value of a car decreases as it gets older. The same thing happens with a company's machines and equipment. But equipment forms part of the company's assets, so it must have an accurate value at any time and the company must know this value. This allowance for this declining value is called **depreciation**.

Most organisations write down their assets each year – which means they reduce the book value of assets. There are several ways of doing this, with the two most widely used being:

- straight-line depreciation

- reducing balance depreciation

These assume that the equipment is bought, used for its expected life, and then sold for its residual – or scrap – value.

## 6.4.1 Straight-line depreciation

Straight-line depreciation reduces the equipment's value by the same amount every year, so:

$$\text{annual depreciation} = \frac{\text{cost of equipment} - \text{residual value}}{\text{estimated life of equipment}}$$

A machine costing £20 000 with a scrap value of £5 000 after a life of 10 years has annual depreciation of:

$$\text{annual depreciation} = \frac{20\,000 - 5000}{10} = £1500 \text{ a year}$$

This means the book value of the machine is £18 500 after one year, £17 000 after two years, and so on. Straight-line depreciation is easy to calculate but it does not show true values. Most equipment loses a lot of value in its first year of operation and will actually be worth less than its depreciated value.

## 6.4.2 Reducing balance depreciation

In the reducing balance method, a fixed percentage of the value of equipment is written off each year. Typically, a piece of equipment has its book value reduced by 20% a year. This gives a more accurate value of equipment, with higher depreciation in the early years.

To find values for the reducing balance method, we use a simple extension of compound interest. We know that an amount $A_0$ that increases at a fixed percentage, $i$, each year has a value after $n$ years of:

$$A_n = A_0 \times (1 + i)^n$$

However, if the amount is decreasing by a fixed percentage – as it does with depreciation – we simply subtract $i$ instead of adding it. Then equipment whose original cost is $A_0$, with a depreciation rate $i$, has a depreciated value, $A_n$, after $n$ years of:

$$A_n = A_0 \times (1 - i)^n$$

## we Worked example 6.7

Penmuir Transport buys a new packing machine for £10 000.

(a) Use the straight-line method to find the annual depreciation if the asset has an expected life of five years and a scrap value of £1000.

(b) If the company uses a reducing balance method with a depreciation rate of 30%, what is the machine's value after five years?

## Solution

(a) For straight-line depreciation we know that:

$$\text{annual depreciation} = \frac{\text{cost of equipment} - \text{scrap value}}{\text{estimated life of equipment}}$$

so:

$$\text{annual depreciation} = \frac{10\,000 - 1000}{5} = £1800$$

(b) For reducing balance depreciation:

$$A_n = A_0 \times (1 - i)^n$$

Here $A_0 = 10\,000$ and $i = 0.3$, so the machine's depreciated value after five years is:

$$A_5 = 10\,000 \times (1 - 0.3)^5 = £1681$$

Figure 6.6 shows these calculations on a spreadsheet with the depreciated value at the end of each year. It also shows how a reducing balance method would need a depreciation rate of 36.9% to give a scrap value of £1000 after five years.

| | A | B | C | D | E | F |
|---|---|---|---|---|---|---|
| 1 | **Depreciation** | | | | | |
| 2 | | | | | | |
| 3 | | | | | | |
| 4 | **Straight-line method** | | | **Reducing balance method** | | |
| 5 | | | | | | |
| 6 | **New value** | 10000 | | **New value** | 10000 | 10000 |
| 7 | **Scrap value** | 1000 | | **Scrap value** | | 1000 |
| 8 | **Useful life** | 5 | | **Useful life** | | 5 |
| 9 | **Annual depreciation** | 1800 | | **Depreciation rate** | 0.3 | 0.369 |
| 10 | | | | | | |
| 11 | **Year** | **Value** | | **Year** | **Value** | **Value** |
| 12 | 0 | 10000 | | 0 | 10000 | 10000 |
| 13 | 1 | 8200 | | 1 | 7000 | 6310 |
| 14 | 2 | 6400 | | 2 | 4900 | 3982 |
| 15 | 3 | 4600 | | 3 | 3430 | 2512 |
| 16 | 4 | 2800 | | 4 | 2401 | 1585 |
| 17 | 5 | 1000 | | 5 | 1681 | 1000 |

**Figure 6.6**  Depreciated values for Worked Example 6.7.

## In summary

**The value of most assets decreases as the assets get older. There are several ways of allowing for this depreciation. The most widely used are the straight-line and reducing balance methods.**

# 6.5 Mortgages and sinking funds

We have seen how the value of an investment rises as it earns interest:

$$A_n = A_0 \times (1 + i)^n$$

Now suppose that we can add an additional investment of $F$ at the end of each period. After $n$ periods the amount invested will have risen to:

$$A_n = A_0 \times (1 + i)^n + \frac{F \times (1 + i)^n - F}{i}$$

In this daunting equation, the first part shows the income from the original investment and the second part shows the amount accumulated by regular payments.

This is often used by organisations that want to set aside regular payments to accumulate a certain amount by the end of some period. This is a **sinking fund** are typically set up to replace equipment or vehicles. A sinking fund does not usually have an initial payment, so $A_0 = 0$.

## we Worked example 6.8

How much should a company put into a sinking fund each year to give £20 000 at the end of ten years when expected interest rates are 15%?

## Solution

We know that the variables are:

$A_n = £20\,000$

$A_0 = £0$

$i = 0.15$

$n = 10$

We can substitute these to give:

$$A_n = A_0 \times (1 + i)^n + \frac{F \times (1 + i)^n - F}{i}$$

so

$$20\,000 = 0 + \frac{F \times (1 + 0.15)^{10} - F}{0.15}$$

$$3000 = (F \times 4.0455) - F$$

that is,

$$F = £985.04$$

The company should put £985.04 into the fund each year, as shown in the spreadsheet in Figure 6.7.

|    | A | B | C | D |
|----|---|---|---|---|
| 1  | **Sinking funds** | | | |
| 2  | | | | |
| 3  | **Initial payment** | | | 0 |
| 4  | **Amount needed** | | | 20000 |
| 5  | **Interest rate** | | | 0.15 |
| 6  | **Annual payment** | | | 985.04 |
| 7  | | | | |
| 8  | | | | |
| 9  | **Year** | **Added at end** | **Interest** | **Value at end** |
| 10 | 1 | 985.04 | 0.00 | 985.04 |
| 11 | 2 | 985.04 | 147.76 | 2117.84 |
| 12 | 3 | 985.04 | 317.68 | 3420.55 |
| 13 | 4 | 985.04 | 513.08 | 4918.67 |
| 14 | 5 | 985.04 | 737.80 | 6641.52 |
| 15 | 6 | 985.04 | 996.23 | 8622.78 |
| 16 | 7 | 985.04 | 1293.42 | 10901.24 |
| 17 | 8 | 985.04 | 1635.19 | 13521.47 |
| 18 | 9 | 985.04 | 2028.22 | 16534.73 |
| 19 | 10 | 985.04 | 2480.21 | 19999.97 |
| 20 | 11 | 985.04 | 3000.00 | 23985.01 |

**Figure 6.7**   Calculations for the sinking fund in Worked
Example 6.8.

You can see that sinking funds are rather like mortgages, but instead of a positive initial payment, mortgages have a negative payment – the amount borrowed – and this is repaid in regular instalments.

## we Worked example 6.9

Summerville & Co. pays for an extension to its premises by a mortgage of £30 000 over 25 years at a fixed interest rate of 12% per annum. It repays the mortgage by regular instalments at the end of every year. How much should each instalment be?

## Solution

We know that:

$$A_0 = -£30,000$$

$$A_{25} = £0$$

$$i = 0.12$$

$$n = 25$$

Now we want to find $F$, and can substitute into the standard equation to get:

$$A_n = A_0 \times (1 + i)^n + \frac{F \times (1 + i)^n - F}{i}$$

$$0 = -30\,000 \times 1.12^{25} + \frac{F \times 1.12^{25} - F}{0.12}$$

Then,

$$30\,000 \times 17 \times 0.12 = (F \times 17) - F$$

or

$$F = £3825$$

After 25 payments of £3826 the original debt will be repaid. Notice that the total payment here is £95 650. It is quite common for total mortgage repayments to be several times the original loan, especially when interest rates are high.

## In summary

**We can do calculations for sinking funds and mortgages. The calculations for these are rather messy but are easy with a spreadsheet.**

## Chapter review

Every organisation is concerned with its finances. There are many quantitative models to help manage finances, and this chapter has described some of the most important ones. In particular, it described::

- break-even points;
- economies of scale;
- the value of money over time with compound interest;
- discounting to present values and internal rates of return;
- depreciation of assets;
- payments of sinking funds and mortgages.

# Problems

6.1 A firm of taxi operators has a fixed cost of £4500 a year for each car. Each kilometre driven costs 20p with fares averaging 30p.
   (a) How many kilometres a year does each car need to travel before making a profit?
   (b) Last year each car drove 160 000 kilometres. What were the incomes and profit?

6.2 An airline is considering a new service between Aberdeen and Calgary. Its existing aeroplanes, each of which has a capacity of 240 passengers, could be used for one flight a week with fixed costs of £30 000 and variable costs amounting to 50% of ticket price. If the airline plans to sell tickets at £200 each, how many passengers will be needed to break even on the proposed route? Does this seem a reasonable number?

6.3 A company is going to introduce a new product and must choose one of three alternatives. Given the data in Table 6.2, which do you think is the best?

**Table 6.2**

|  | Product | | |
|---|---|---|---|
|  | A | B | C |
| Annual sales | 600 | 900 | 1200 |
| Unit cost | 680 | 900 | 1200 |
| Fixed cost | 200 000 | 350 000 | 500 000 |
| Product life | 3 years | 5 years | 8 years |
| Selling price | 760 | 1000 | 1290 |

6.4 How much will an initial investment of £1000 earning interest of 8% a year be worth at the end of twenty years?

6.5 Several years ago a couple bought an endowment insurance policy which has recently matured. They have the option of receiving £20 000 now or £40 000 in ten years' time. Because they have retired and pay no income tax, they could invest the money with a real interest rate of 10% a year for the foreseeable future. Which option should they take?

6.6 The cash flows for three projects are given in Table 6.3. Calculate the net present value using a discounting rate of 12% a year. What are the internal rates of return?

**Table 6.3**

|  | Project A | | Project B | | Project C | |
|---|---|---|---|---|---|---|
| Year | Income | Expenditure | Income | Expenditure | Income | Expenditure |
| 0 | 0 | 18 000 | 0 | 24 000 | 0 | 21 000 |
| 1 | 2500 | 0 | 2000 | 10 000 | 0 | 12 000 |
| 2 | 13 500 | 6000 | 10 000 | 6000 | 20 000 | 5000 |
| 3 | 18 000 | 0 | 20 000 | 2000 | 20 000 | 1000 |
| 4 | 6000 | 2000 | 30 000 | 2000 | 30 000 | 0 |
| 5 | 1000 | 0 | 30 000 | 2000 | 30 000 | 0 |

6.7    What is the internal rate of return for a product which gives the following net cash flow?

| Year | 1 | 2 | 3 | 4 | 5 | 6 | 7 | 8 | 9 |
|------|---|---|---|---|---|---|---|---|---|
| Net cash flow | −6000 | −1500 | −500 | 600 | 1800 | 2000 | 1400 | 300 | 100 |

6.8    A company buys new vehicles for £15 000. If these have an expected life of six years, and a scrap value of £2000, use the straight-line method to find the annual depreciation. If the reducing balance method is used with a depreciation rate of 25%, what is the value of the vehicles after six years?

6.9    A company makes fixed annual payments to a sinking fund to replace equipment in five years' time. The equipment is valued at £100 000 and interest rates are 12%. How much should each payment be?

6.10   Five projects generate the incomes shown in Table 6.4 (in thousands of pounds) over the next ten years. Which project do you think is most attractive?

**Table 6.4**

| Project | A | B | C | D | E |
|---------|------|------|-----|------|------|
| Year 1 | −120 | −200 | −60 | 0 | −500 |
| Year 2 | −60 | 0 | 30 | 10 | −200 |
| Year 3 | 5 | 100 | 30 | 20 | −100 |
| Year 4 | 30 | 80 | 30 | 30 | 50 |
| Year 5 | 45 | 60 | 30 | 30 | 100 |
| Year 6 | 55 | 50 | −40 | 20 | 200 |
| Year 7 | 65 | 40 | 30 | −100 | 300 |
| Year 8 | 65 | 40 | 30 | 50 | 350 |
| Year 9 | 60 | 35 | 30 | 40 | 400 |
| Year 10 | 50 | 35 | 30 | 30 | 450 |

## we Case study – Mrs Hamilton's retirement savings

Mrs Hamilton has just had her 55th birthday. For many years she has been investing in endowment insurance policies and some of these have now matured to give her a lump sum of around £50 000. She is self-employed and plans to continue working until she is 65. This means that Mrs Hamilton is looking for an investment which will increase in value over the next ten years.

Mrs Hamilton went to her bank manager for some advice. When she suggested that she could add an additional £2 000 a year to her savings, the manager did some sums, and made several suggestions:

- A Savings Account which would give a return of 7.5% a year.

- A Gold Account for the fixed sum. This would give her a return of 9% but leaves the money tied up for at least a year. The additional savings could go into a Savings Account.

- A Personal Accumulator which gives 5% interest on a minimum of £30 000, but 15% on any extra savings that are made.

Mrs Hamilton also visited a building society manager who gave her similar advice, but suggested an additional option:

> 'The most secure investment is to put the money in our Inflation Fighter. This has an interest rate which is linked to the retail price index, and is guaranteed to give a return that is 1% above inflation.'

The building society manager also discussed the possibility of buying a house as an investment.

> 'The housing market has been very unsettled lately. If you take a long-term view, house prices rise by an average of 5% to 15% a year, so houses are a good investment. They can also generate income from rent. Usually you could expect to go about 0.5% of the value of the house a month Something like a quarter of this is needed for repairs and maintenance. Some of my customers take out a mortgage with interest of 10% a year to add to their savings and buy a bigger house.'

> Mrs Hamilton went home and did some thinking. She could calculate how much each of the the the investments would be worth at the end of ten years. Perhaps, though, she should decide how much she wants to have when she retires, and add to her savings to achieve this amount. She thought that £150 000 would be enough, but maybe she should aim for a larger amount. She wants a reasonable income over the next twenty years, even when taking inflation into account. Maybe an annuity would suit her. What help could you give?

# 7 Relating variables by regression

## Contents

## Chapter outline

This chapter shows how regression can find the relationship between variables. After reading the chapter you should be able to:

- discuss the noise in a relationship
- find the line of best fit using linear regression
- use linear regression to make forecasts
- calculate coefficients of determination and correlation
- understand how multiple regression works
- outline the use of non-linear regression

# 7.1 Noise and errors

Last spring, Gateshead Food Processors kept a record of the average daily temperature and the amount of electricity they used.

| Observation | 1 | 2 | 3 | 4 | 5 | 6 |
|---|---|---|---|---|---|---|
| Temperature | 5 | 7 | 10 | 12 | 15 | 17 |
| Electricity | 17 | 21 | 24 | 26 | 32 | 40 |

Figure 7.1 shows that there is a clear linear relationship between temperature and electricity use. But this is not a perfect relationship, as there is some variation about the line. In other words, there is an underlying linear trend, and superimposed on this are some random variations. These variations are called **noise**.

actual value = underlying pattern + random noise

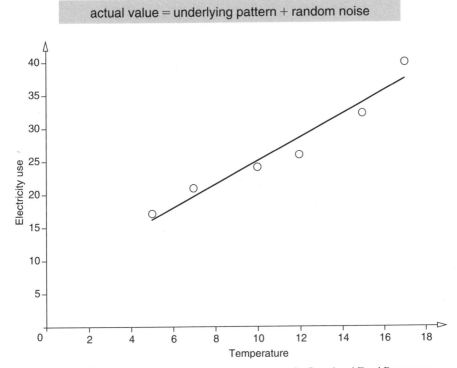

**Figure 7.1**  Graph of temperature against electricity use in Gateshead Food Processors.

The amount of noise shows how strong a relationship is.

- If there is no noise, the relationship is perfect.
- If there is some noise the relationship becomes weaker.
- If there is a lot of noise, the relationship is very weak.

So we need a way of measuring the noise. The best way of doing this is to imagine the noise as an error in each observation. This error is the difference between the actual value and the value expected from the relationship.

$E_i$ = actual value of observation $i$ – value expected from the relationship

We can combine these separate errors into a single measure to show the strength of the relationship. We could use the mean error, but we saw in Chapter 5 that this allowed positive and negative errors to cancel. So again we are going to use the mean squared error.

## we Worked example 7.1

Elma Brown thinks that the following values are related by the equation $y = 3x + 3$. If she is right, what is the mean squared error?

| Observation | 1 | 2 | 3 | 4 | 5 | 6 |
|---|---|---|---|---|---|---|
| $x$ | 3 | 6 | 10 | 8 | 4 | 1 |
| $y$ | 10 | 24 | 29 | 25 | 12 | 5 |

## Solution

We can find the expected values of $y$ by substituting in the equation $y = 3x + 3$. Then the error in each observation comes from:

$E_i$ = actual $y_i$ – expected $y_i$

| Observation | 1 | 2 | 3 | 4 | 5 | 6 |
|---|---|---|---|---|---|---|
| $x$ | 3 | 6 | 10 | 8 | 4 | 1 |
| actual $y$ | 10 | 24 | 29 | 25 | 12 | 5 |
| expected $y$ | 9 | 21 | 33 | 27 | 15 | 6 |
| error, $E_i$ | 1 | 3 | −4 | −2 | −3 | −1 |

Now we can calculate the mean squared error:

$$\text{mean squared error} = \frac{1^2 + 3^2 + (-4)^2 + (-2)^2 + (-3)^2 + (-1)^2}{6}$$

$$= \frac{1 + 9 + 16 + 4 + 9 + 1}{6} = 6.67$$

You can find many examples of relationships with superimposed noise. The sales of a product depend on the price being charged, demand for a service depends on advertising expenditure, productivity depends on bonus payments, borrowing depends on interest rates, crop size depends on the amount of fertilizer used, and so on. These are examples of causal relationships where changes in the first (dependent) variable are actually caused by changes in the second (independent) variable.

Nevertheless, even when there is a strong relationship between variables, you must not assume that there is any cause and effect. There is a clear relationship between the sales of ice cream and the sales of sunglasses, but there is no cause and effect here – and you cannot increase sales of ice cream by increasing the sales of sunglasses. You can find many ridiculous examples of this mistake: the number of prosecutions for drunken driving in an area is related to the number of lamp-posts; the birth rate in the UK is related to the number of storks nesting in Sweden; the number of people in higher education is related to life expectancy, and so on.

## In summary

**The amount of noise shows the strength of a relationship. We can measure the noise by the mean squared error. Even when there is a strong relationship, there need not be any cause and effect.**

# 7.2 Finding the line of best fit

To find a linear relationship between two variables, we are going to use linear regression. The following example shows the overall approach.

## we Worked example 7.2

The following table shows the number of shifts worked each month in a factory, and the output. If 30 shifts are planned for next month, what is the expected output?

| Month | 1 | 2 | 3 | 4 | 5 | 6 | 7 | 8 | 9 |
|---|---|---|---|---|---|---|---|---|---|
| Shifts worked | 50 | 70 | 25 | 55 | 20 | 60 | 40 | 25 | 35 |
| Output | 352 | 555 | 207 | 508 | 48 | 498 | 310 | 153 | 264 |

## Solution

The best thing to do with a set of data is to draw a graph. Figure 7.2 shows a scatter diagram of shifts worked (the independent variable, $x$) and units made (the dependent variable, $y$). There is a clear linear relationship, which we can show by drawing a straight line through the data. From this line, you can see that working 30 shifts gives an output of around 200 units.

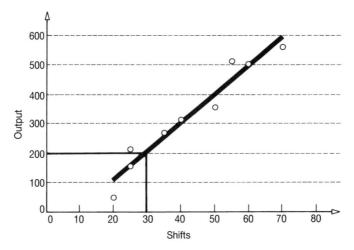

**Figure 7.2**   Linear relationship on a scatter diagram.

In Worked Example 7.2, we drew a scatter diagram, noticed a linear relationship and drew a line of best fit by eye. However, it would clearly be better to calculate a **line of best fit**.

The equation of a straight line is:

$$y = a + bx$$

where:

$x$ = independent variable

$y$ = dependent variable

$a$ = point where the line intersects the $y$-axis

$b$ = gradient of the line

To get the line of best fit, we have to find the values of the constants $a$ and $b$. Linear regression does this by finding the values that minimize the mean squared error. It does this using the standard result:

$$b = \frac{n\Sigma xy - \Sigma x \Sigma y}{n\Sigma x^2 - (\Sigma x)^2}$$

$$a = \bar{y} - b\bar{x}$$

Where $\bar{x}$ and $\bar{y}$ are the mean values and $n$ is the number of observations.

# we Worked example 7.3

Find the line of best fit through the following data for an advertising budget (in thousands of pounds) and units sold. How many units will be sold if the advertising budget is £30 000?

| Month | 1 | 2 | 3 | 4 |
|---|---|---|---|---|
| Advertising budget, x | 20 | 40 | 60 | 80 |
| Units sold, y | 120 | 170 | 210 | 230 |

## Solution

Figure 7.3 shows a clear linear relationship with:

units sold $(y) = a + b \times$ advertising budget $(x)$

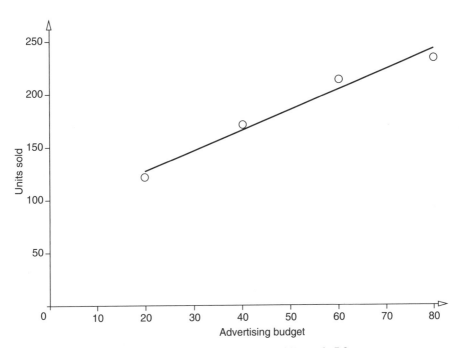

**Figure 7.3**   Linear trend in Worked Example 7.3.

The top part of Figure 7.4 shows the linear regression calculations on a spreadsheet, where the computer finds:

$$n = 4$$

$$\bar{x} = \frac{200}{4} = 50$$

$$\bar{y} = \frac{730}{4} = 182.5$$

$$b = \frac{n\Sigma xy - \Sigma x \Sigma y}{n\Sigma x^2 - (\Sigma x)^2}$$

$$= \frac{(4 \times 40\,200) - (200 \times 730)}{(4 \times 12\,000) - (200 \times 200)} = 1.85$$

$$a = \bar{y} - b\bar{x} = 182.5 - (1.85 \times 50) = 90$$

The line of best fit, $y = a + bx$, is:

units sold $= 90 + (1.85 \times$ advertising budget$)$

With an advertising budget of £30 000:

units sold $= 90 + (1.85 \times 30) = 146$

Spreadsheets have a standard function for doing regression, illustrated in the second part of Figure 7.4. The computer calculates some regression statistics – which we talk about in the next section – then it prints results for the 'intercept' and 'X variable 1', which are the two figures we want – 90 and 1.85.

|    | A | B | C | D | E | F |
|----|---|---|---|---|---|---|
| 1 | **Linear regression** | | | | | |
| 2 | | | | | | |
| 3 | **Doing calculation** | | | | | |
| 4 | | | | | | |
| 5 | | **Data** | | | **Calculations** | |
| 6 | **Month** | **x** | **y** | | **xy** | **x^2** |
| 7 | 1 | 20 | 120 | | 2400 | 400 |
| 8 | 2 | 40 | 170 | | 6800 | 1600 |
| 9 | 3 | 60 | 210 | | 12600 | 3600 |
| 10 | 4 | 80 | 230 | | 18400 | 6400 |
| 11 | **Sums** | **200** | **730** | | **40200** | **12000** |
| 12 | **Means** | **50** | **182.5** | | | |
| 13 | | | | | | |
| 14 | **Substitution** | | **b** | **1.85** | **a** | **90** |
| 15 | | | | | | |
| 16 | **Using standard function** | | | | | |
| 17 | | | | | | |
| 18 | *Regression statistics* | | | | | |
| 19 | Correlation, r | 0.984 | | | | |
| 20 | Determination r^2 | 0.967 | | | | |
| 21 | Adjusted r^2 | 0.951 | | | | |
| 22 | Standard error | 10.724 | | | | |
| 23 | Observations | 4 | | | | |
| 24 | | | | | | |
| 25 | | | | | | |
| 26 | *Coefficients* | | | | | |
| 27 | Intercept | 90 | | | | |
| 28 | X variable 1 | 1.85 | | | | |

**Figure 7.4**   Finding the line of best fit in Worked Example 7.3.

## we Worked example 7.4

A company is about to change the way it inspects a product. It did some experiments with differing numbers of inspections and found the following average number of defects.

| Inspections | 0 | 1 | 2 | 3 | 4 | 5 | 6 | 7 | 8 | 9 | 10 |
|-------------|---|---|---|---|---|---|---|---|---|---|----|
| Defects | 92 | 86 | 81 | 72 | 67 | 59 | 53 | 43 | 32 | 24 | 12 |

If the company makes six inspections, how many defects would it expect to find? What is the effect of making 20 inspections?

## Solution

Figure 7.5 shows a linear relationship, with $x$ as the number of inspections and $y$ as the number of defects.

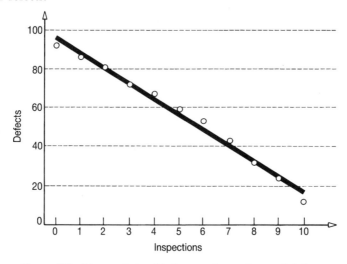

**Figure 7.5**   Linear relationship between inspections and defects.

You can use a computer to find the line of best fit, or the calculations:

$$b = \frac{n\Sigma(xy) - \Sigma x \Sigma y}{n\Sigma x^2 - (\Sigma x)^2}$$

$$= \frac{(11 \times 2238) - (55 \times 621)}{(11 \times 385) - (55 \times 55)} = -7.88$$

$$a = \bar{y} - b\bar{x} = \frac{621}{11} + 7.88 \times \frac{55}{11} = 95.85$$

The line of best fit is:

$$y = 95.85 - 7.88x$$

or

defects $= 95.85 - 7.88 \times$ number of inspections

With six inspections the company would expect

$95.85 - 7.88 \times 6 = 48.57$ defects

With 20 inspections we have to be a little more careful as substitution gives

$95.85 - 7.88 \times 20 = -61.75$.

It is impossible to have a negative number of defects, so we would simply forecast zero defects.

## we Worked example 7.5

Demand for a product over the past eight weeks has been 17, 23, 41, 38, 42, 47, 51 and 56. Use linear regression to forecast demand for the next two weeks.

## Solution

This example shows how to use linear regression for forecasting, by putting the week number as the independent variable and demand as the dependent variable.

| Week, $x$ | 1 | 2 | 3 | 4 | 5 | 6 | 7 | 8 |
|-----------|----|----|----|----|----|----|----|----|
| Demand, $y$ | 17 | 23 | 41 | 38 | 42 | 47 | 51 | 56 |

Solving the regression equations in the usual way gives the line of best fit as:

demand $= 16.07 + 5.18 \times$ week

Then

week $= 9$:    demand $= 16.07 + (5.18 \times 9) = 62.69$

week $= 10$:    demand $= 16.07 + (5.18 \times 10) = 67.87$

### In summary

**Linear regression finds the line of best fit through a set of data. This is the line that minimizes the sum of squared errors.**

# 7.3 Measuring the strength of a relationship

## 7.3.1 Coefficient of determination

Now we can find the line of best fit through a set of data, but still need some way of measuring how good this line is. If the errors are small then the line is a good fit; but if the errors are large even the best line is not very good. To measure the goodness of fit we will use a measure called the **coefficient of determination**.

Imagine a set of data, like the one shown in Figure 7.6. This has a mean value of $\bar{y}$. An observation $y$ has a variation from this mean. Some of this variation is explained by the linear relationship, and some is due to noise and is unexplained.

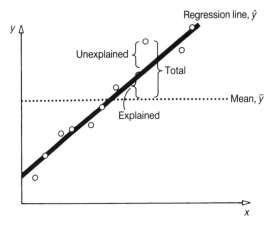

**Figure 7.6**   Relationships between total, explained and
unexplained variations.

If most of the total error is explained by the linear relationship, this shows a good fit.
So we can define:

$$\text{coefficient of determination} = \frac{\text{explained sum of squared error}}{\text{total sum of squared error}}$$

The coefficient of determination has a value between 0 and 1. If it is close to 1, most of
the variation is explained by the relationship, and the line is a good fit. If the value is
close to 0, most of the variation is unexplained and the line is not a good fit.

The equation for calculating the coefficient of determination is rather messy.

$$\text{coefficient of determination} = \left[ \frac{n\Sigma xy - \Sigma x\Sigma y}{\sqrt{\{n\Sigma x^2 - (\Sigma x)^2\} \times \{n\Sigma y^2 - (\Sigma y)^2\}}} \right]^2$$

## we Worked example 7.6

Calculate the coefficient of determination for the data in Worked Example 7.4.

## Solution

We have already done the regression for these figures and found that the line of best fit
is $y = 95.85 - 7.88x$. Now we are checking to see how good this line is. The values used
in calculating this are set out in Table 7.2.

**Table 7.2**

|       | 0    | 1    | 2    | 3    | 4    | 5    | 6    | 7    | 8    | 9   | 10  | Totals |
|-------|------|------|------|------|------|------|------|------|------|-----|-----|--------|
| $x$   | 0    | 1    | 2    | 3    | 4    | 5    | 6    | 7    | 8    | 9   | 10  | 55     |
| $y$   | 92   | 86   | 81   | 72   | 67   | 59   | 53   | 43   | 32   | 24  | 12  | 621    |
| $xy$  | 0    | 86   | 162  | 216  | 268  | 295  | 318  | 301  | 256  | 216 | 120 | 2238   |
| $x^2$ | 0    | 1    | 4    | 9    | 16   | 25   | 36   | 49   | 64   | 81  | 100 | 385    |
| $y^2$ | 8464 | 7396 | 6561 | 5184 | 4489 | 3481 | 2809 | 1849 | 1024 | 576 | 144 | 41977  |

Now we can substitute these values to find the coefficient of determination.

$$\text{coefficient of determination} = \left[ \frac{n\Sigma xy - \Sigma x\Sigma y}{\sqrt{\{n\Sigma x^2 - (\Sigma x)^2\} \times \{n\Sigma y^2 - (\Sigma y)^2\}}} \right]^2$$

$$= \left[ \frac{(11 \times 2238) - (55 \times 621)}{\sqrt{\{(11 \times 385) - (55 \times 55)\} \times \{(11 \times 41\,977) - (621 \times 621)\}}} \right]^2$$

$$= -0.9938^2$$

$$= 0.9877$$

This shows that 98.77% of the variation can be explained by the linear relationship and only 1.23% is unexplained. The regression line is obviously a very good fit to the data. Normally, any value for the coefficient of determination above about 0.5 is considered a reasonable fit.

## In summary

**The coefficient of determination measures the proportion of the total variation explained by the regression line. A value close to 1 shows a good fit of the regression line, while a value close to 0 shows a poor fit.**

## 7.3.2 Coefficient of correlation

Another useful measure in regression is the **coefficient of correlation**. This answers the basic question, are $x$ and $y$ linearly related? The coefficients of correlation and determination answer very similar questions, and in practice:

**coefficient of correlation** $= \sqrt{\text{coefficient of determination}}$

The coefficient of determination has a value between 0 and 1, and is usually called $r^2$. The coefficient of correlation has a value between $-1$ and $+1$ and is usually called $r$.

- A value of $r = 1$ shows a perfect linear relationship with no noise at all, and as one variable increases so does the other.

- A positive value of $r$ close to zero shows a weak linear relationship.

- A value of $r = 0$ shows that there is no linear relationship between the two variables.

- A negative value of $r$ close to zero shows a weak linear relationship.

- A value of $r = -1$ shows that the two variables have a perfect linear relationship and as one increases the other decreases.

These results are shown in Figure 7.7.

When $r$ is between 0.7 and $-0.7$ the coefficient of determination, $r^2$, is less than 0.49. Less than half the variation is explained by the linear relationship, so the results are not very reliable.

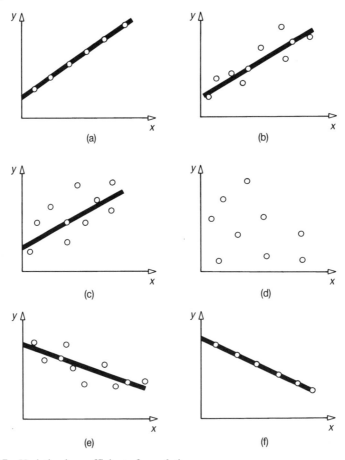

**Figure 7.7** Variation in coefficient of correlation:
(a) $r = +1$ (perfect positive correlation); (b) $r$ is close to $+1$ (line is good fit);
(c) $r$ is decreasing (line is poor fit); (d) $r = 0$ (random points);
(e) $r$ is close to $-1$ (line is good fit); (f) $r = -1$ (perfect negative correlation).

## we Worked example 7.7

Calculate the coefficients of correlation and determination for the following data. What conclusions can you reach from these? What is the line of best fit?

| x | 4 | 17 | 3 | 21 | 10 | 8 | 4 | 9 | 13 | 12 | 2 | 6 | 15 | 8 | 19 |
|---|---|----|---|----|----|---|---|---|----|----|---|---|----|---|----|
| y | 13 | 47 | 24 | 41 | 29 | 33 | 28 | 38 | 46 | 32 | 14 | 22 | 26 | 21 | 50 |

## Solution

Figure 7.8 on the following page shows the calculations on a spreadsheet.

You can check the calculations to find the coefficient of correlation, $r$:

$$r = \frac{n\Sigma(xy) - \Sigma x \Sigma y}{\sqrt{\{n\Sigma x^2 - (\Sigma x)^2\} \times \{n\Sigma y^2 - (\Sigma y)^2\}}}$$

$$= \frac{15 \times 5442 - 151 \times 464}{\sqrt{(15 \times 2019 - 151 \times 151) \times (15 \times 16230 - 464 \times 464)}}$$

$$= 0.797$$

Squaring this gives the coefficient of determination:

$$r^2 = 0.635$$

This shows that 63.5% of the variation is explained by the linear relationship, and only 26.5% is unexplained. This is quite a strong linear relationship, which means that the line of best fit is fairly reliable. The line of best fit is given by the equation:

$$y = 15.376 + 1.545x$$

| | A | B | C | D | E | F | G | H |
|---|---|---|---|---|---|---|---|---|
| 1 | **Linear regression** | | | | | | | |
| 2 | | | | | | | | |
| 3 | **Doing calculation** | | | | | | | |
| 4 | | | | | | | | |
| 5 | | **Data** | | | **Calculations** | | | |
| 6 | **Observation** | **x** | **y** | | **xy** | **x^2** | **y^2** | |
| 7 | 1 | 4 | 13 | | 52 | 16 | 169 | |
| 8 | 2 | 17 | 47 | | 799 | 289 | 2209 | |
| 9 | 3 | 3 | 24 | | 72 | 9 | 576 | |
| 10 | 4 | 21 | 41 | | 861 | 441 | 1681 | |
| 11 | 5 | 10 | 29 | | 290 | 100 | 841 | |
| 12 | 6 | 8 | 33 | | 264 | 64 | 1089 | |
| 13 | 7 | 4 | 28 | | 112 | 16 | 784 | |
| 14 | 8 | 9 | 38 | | 342 | 81 | 1444 | |
| 15 | 9 | 13 | 46 | | 598 | 169 | 2116 | |
| 16 | 10 | 12 | 32 | | 384 | 144 | 1024 | |
| 17 | 11 | 2 | 14 | | 28 | 4 | 196 | |
| 18 | 12 | 6 | 22 | | 132 | 36 | 484 | |
| 19 | 13 | 15 | 26 | | 390 | 225 | 676 | |
| 20 | 14 | 8 | 21 | | 168 | 64 | 441 | |
| 21 | 15 | 19 | 50 | | 950 | 361 | 2500 | |
| 22 | **Sums** | **151** | **464** | | **5442** | **2019** | **16230** | |
| 23 | **Means** | **10.067** | **30.933** | | | | | |
| 24 | | | | | | | | |
| 25 | **Substitution** | | **a** | **15.376** | **b** | **1.545** | | |
| 26 | | | **r** | **0.797** | **r^2** | **0.635** | | |
| 27 | | | | | | | | |
| 28 | **Using standard function** | | | | | | | |
| 29 | | | | | | | | |
| 30 | *Regression statistics* | | | | | | | |
| 31 | Correlation, r | 0.797 | | | | | | |
| 32 | Determination r^2 | 0.635 | | | | | | |
| 33 | Adjusted r^2 | 0.607 | | | | | | |
| 34 | Standard error | 7.261 | | | | | | |
| 35 | Observations | 15 | | | | | | |
| 36 | | | | | | | | |
| 37 | | | | | | | | |
| 38 | *Coefficients* | | | | | | | |
| 39 | Intercept | 15.376 | | | | | | |
| 40 | X variable 1 | 1.545 | | | | | | |

**Figure 7.8**  Calculations for Worked Example 7.7.

## In summary

**The correlation coefficient shows how strong the linear relationship is between two variables. A value close to $+1$ or $-1$ shows a strong relationship, while a value close to zero shows a weak one.**

# 7.4 Extensions to linear regression

## 7.4.1 Multiple regression

There are several extensions to the basic linear regression model. Here we will describe the two most important: multiple regression and non-linear regression.

Linear regression looked at problems where $y = a + bx$ with, for example, sales of a product related to the amount spent on advertising:

$$sales = a + b \times advertising$$

Suppose, though, that sales were also related to price, unemployment rates, average incomes, competition, and so on. In other words, the dependent variable, $y$, is not set by a single independent variable, $x$, but by a number of separate independent variables $x_i$. Then we could write:

$$y = a + b_1 x_1 + b_2 x_2 + b_3 x_3 + b_4 x_4 + b_5 x_5 + \ldots$$

or in this example:

$$sales = a + (b_1 \times advertising) + (b_2 \times price) + (b_3 \text{ unemployment rate})$$
$$+ (b_4 \times income) + (b_5 \times competition) + \ldots$$

By using more independent variables we are trying to get a better model for describing sales. We might find, for example, that advertising explains 60% of the variation in sales, but that adding another term for price explains 75% of the variation, and adding another term for unemployment explains 85% of the variation, and so on.

We are looking for a linear relationship between a dependent variable and a set of independent ones, so we should really call this approach **multiple linear regression**, but it is usually shortened to **multiple regression**. Now we need a way of calculating the constants $a$ and $b_i$. As you can imagine, these are rather complicated, but a lot of software has standard functions for multiple regression.

## We Worked example 7.8

The spreadsheet in Figure 7.9 shows the sales, advertising cost and price for one of Causehead Trader's products. The rest of the figure gives some results for multiple regression. What do these results show?

| | A | B | C |
|---|---|---|---|
| 1 | **Multiple regression** | | |
| 2 | | | |
| 3 | **Data** | | |
| 4 | **Sales** | **Advertising** | **Price** |
| 5 | 2450 | 100 | 50 |
| 6 | 3010 | 130 | 56 |
| 7 | 3090 | 160 | 45 |
| 8 | 3700 | 190 | 63 |
| 9 | 3550 | 210 | 48 |
| 10 | 4280 | 240 | 70 |
| 11 | | | |
| 12 | **Results** | | |
| 13 | | | |
| 14 | *Regression statistics* | | |
| 15 | Correlation, r | 0.9958 | |
| 16 | Determination r^2 | 0.9916 | |
| 17 | Adjusted r^2 | 0.9861 | |
| 18 | Standard error | 75.0553 | |
| 19 | Observations | 6 | |
| 20 | | | |
| 21 | *Coefficients* | | |
| 22 | Intercept | 585.965 | |
| 23 | X variable 1 | 9.923 | |
| 24 | X variable 2 | 19.107 | |

**Figure 7.9**   Spreadsheet for multiple regression in Worked Example 7.8.

## Solution

We are given data for two independent variables – advertising and price – and one dependent variable – sales. So we are looking for a relationship:

$$\text{sales} = a + (b_1 \times \text{advertising}) + (b_2 \times \text{price})$$

Lines 22–24 of Figure 7.9 show the line of best fit as:

$$\text{sales} = 585.965 + (9.923 \times \text{advertising}) + (19.107 \times \text{price})$$

The coefficient of correlation, $r$, is 0.9958, which shows a very strong linear relationship. The coefficient of determination, $r^2$, shows that 99.16% of the variation is explained by the relationship. You can see on line 17 an 'adjusted $r^2$'. Sometimes the normal value is a bit optimistic, especially when there are only a few observations. To overcome this bias, the computer calculates an adjusted value.

We have to take a few precautions when dealing with multiple regression. For example, the method only works properly if there is no significant linear relationship between the independent variables. So in Worked Example 7.8 there should be no relationship between the advertising costs and price. In practice, we can accept the results if these relationships are not too strong.

The coefficients of correlation show how strong the relationships are between independent variables. Figure 7.10 shows the coefficients of correlation from Worked Example 7.9. This shows, not surprisingly, a perfect correlation between each variable and itself. We want a very high correlation between sales and each of the independent variables – as shown in the values of 0.9653 between sales and advertising and 0.7228 between sales and price. We also want a very low correlation between advertising and price. The value of 0.5348 is a bit high, but the result is not too bad.

| | A | B | C | D |
|---|---|---|---|---|
| 1 | **Multiple regression** | | | |
| 2 | | | | |
| 3 | **Data** | | | |
| 4 | **Sales** | **Advertising** | **Price** | |
| 5 | 2450 | 100 | 50 | |
| 6 | 3010 | 130 | 56 | |
| 7 | 3090 | 160 | 45 | |
| 8 | 3700 | 190 | 63 | |
| 9 | 3550 | 210 | 48 | |
| 10 | 4280 | 240 | 70 | |
| 11 | | | | |
| 12 | **Correlation** | | | |
| 13 | | Sales | Advertising | Price |
| 14 | Sales | 1 | | |
| 15 | Advertising | 0.9653 | 1 | |
| 16 | Price | 0.7228 | 0.5348 | 1 |

**Figure 7.10**  Coefficients of correlation for variables in Worked Example 7.8.

A simple linear regression model relating sales to advertising explains 93.18% of the variation in sales. This comes from the coefficient of determination, $r^2$, for a coefficient of correlation, $r$, of 0.9653. But when we add price as a second variable, we increase this to 99.16% (shown in line 16 of Figure 7.9).

## we Worked example 7.9

Elsom Service Corporation wants to see how the number of shifts worked, bonus rates paid to employees, average hours of overtime and staff morale affect production. The company has collected the data shown in Table 7.3, using some consistent units. What conclusions can Elsom reach from these data?

**Table 7.3**

| Production | 2810 | 2620 | 3080 | 4200 | 1500 | 3160 | 4680 | 2330 | 1780 | 3910 |
|---|---|---|---|---|---|---|---|---|---|---|
| Shifts | 6 | 3 | 3 | 4 | 1 | 2 | 2 | 7 | 1 | 8 |
| Bonus | 15 | 20 | 5 | 5 | 7 | 12 | 25 | 10 | 12 | 3 |
| Overtime | 8 | 10 | 22 | 31 | 9 | 22 | 30 | 5 | 7 | 20 |
| Morale | 5 | 6 | 3 | 2 | 8 | 10 | 7 | 7 | 5 | 3 |

## Solution

Figure 7.11 shows some calculations for this problem.

| | A | B | C | D | E | F |
|---|---|---|---|---|---|---|
| 1 | **Multiple regression** | | | | | |
| 2 | | | | | | |
| 3 | **Data** | | | | | |
| 4 | **Production** | **Shifts** | **Bonus** | **Overtime** | **Morale** | |
| 5 | 2810 | 6 | 15 | 8 | 5 | |
| 6 | 2620 | 3 | 20 | 10 | 6 | |
| 7 | 3080 | 3 | 5 | 22 | 3 | |
| 8 | 4200 | 4 | 5 | 31 | 2 | |
| 9 | 1500 | 1 | 7 | 9 | 8 | |
| 10 | 3160 | 2 | 12 | 22 | 10 | |
| 11 | 4680 | 2 | 25 | 30 | 7 | |
| 12 | 2330 | 7 | 10 | 5 | 7 | |
| 13 | 1780 | 1 | 12 | 7 | 5 | |
| 14 | 3910 | 8 | 3 | 20 | 3 | |
| 15 | | | | | | |
| 16 | **Correlation** | | | | | |
| 17 | | Production | Shifts | Bonus | Overtime | Morale |
| 18 | Production | 1.0000 | | | | |
| 19 | Shifts | 0.2623 | 1.0000 | | | |
| 20 | Bonus | 0.1531 | −0.3147 | 1.0000 | | |
| 21 | Overtime | 0.8778 | −0.1085 | −0.0220 | 1.0000 | |
| 22 | Morale | −0.3321 | −0.3947 | 0.4512 | −0.2654 | 1.0000 |
| 23 | | | | | | |
| 24 | **Regression** | | | | | |
| 25 | *Regression statistics* | | | | *Coefficient* | |
| 26 | Correlation, r | 0.997 | | Intercept | 346.331 | |
| 27 | Determination r^2 | 0.995 | | X variable 1 | 181.805 | |
| 28 | Adjusted r^2 | 0.990 | | X variable 2 | 50.132 | |
| 29 | Standard error | 100.807 | | X variable 3 | 96.172 | |
| 30 | Observations | 10 | | X variable 4 | −28.705 | |

**Figure 7.11**  Multiple regression results for Worked Example 7.9.

Lines 26–30 show the line of best fit as:

$$\text{production} = 346.331 + (181.805 \times \text{shifts}) + (50.132 \times \text{bonus})$$
$$+ (96.172 \times \text{overtime}) - (28.705 \times \text{morale})$$

This model fits the data very well: the coefficients of correlation and determination are 0.997 and 0.995 respectively. This means that only 0.5% of the variation is unexplained noise. The coefficient of correlation for each pair of independent variables is low, so these pairs are not strongly related. However, we would like higher coefficients of correlation between production and the independent variables. Apart from the correlation between production and overtime, these values are very low – and there is surprising small negative correlation between production and morale.

Elsom should look at its data carefully and see if it can improve the model. Some of the statistical analyses described later in the book can help with this.

### In summary

**Sometimes a dependent variable is related to a number of independent variables. We can use multiple regression to find the values of $a$ and $b_i$ which define the best relationship. Many packages have standard functions for multiple regression.**

### 7.4.2 Non-linear regression

Sometimes there is a clear pattern in the data but it is not linear. There might, for example, be an exponential growth or a quadratic relationship. Then we want to fit the best curve through the data. This is called **non-linear regression**. In principle, the approach is the same as linear regression, but the arithmetic is more complicated.

## we Worked example 7.10

Helen Stanmore is convinced that her accountant has raised his prices more than the cost of inflation. Over the past eleven years she has noticed that the cost of doing her accounts (in thousands of pounds) has been as follows. What do these data show?

| Year | 1 | 2 | 3 | 4 | 5 | 6 | 7 | 8 | 9 | 10 | 11 |
|------|-----|-----|-----|-----|-----|-----|-----|-----|-----|-----|-----|
| Cost | 0.8 | 1.0 | 1.3 | 1.7 | 2.0 | 2.4 | 2.9 | 3.8 | 4.7 | 6.2 | 7.5 |

## Solution

These data are clearly not a straight line. One option is to draw a 'growth' curve through the data. This curve has the standard form, $y = bm^x$, where $x$ and $y$ are the independent and dependent variables, and $b$ and $m$ are constants. Many packages will find the best curve automatically, as shown in the spreadsheet of Figure 7.12.

| | A | B | C | D | E | F | G | H |
|---|---|---|---|---|---|---|---|---|
| 1 | **Non-linear regression** | | | | | | | |
| 2 | | | | | | | | |
| 3 | **Data** | | | **Results** | | **Equation** | | |
| 4 | **Year** | **Price** | | **Predicted price** | | **Coefficients** | | |
| 5 | 1 | 0.8 | | 0.8160 | | **y = bm^x** | | |
| 6 | 2 | 1 | | 1.0179 | | **b** | 0.6541 | |
| 7 | 3 | 1.3 | | 1.2699 | | **m** | 1.2475 | |
| 8 | 4 | 1.7 | | 1.5842 | | | | |
| 9 | 5 | 2 | | 1.9763 | | | | |
| 10 | 6 | 2.4 | | 2.4654 | | | | |
| 11 | 7 | 2.9 | | 3.0756 | | | | |
| 12 | 8 | 3.8 | | 3.8368 | | | | |
| 13 | 9 | 4.7 | | 4.7864 | | | | |
| 14 | 10 | 6.2 | | 5.9710 | | | | |
| 15 | 11 | 7.5 | | 7.4489 | | | | |
| 16 | | | | | | | | |
| 17 | | | | | | | | |
| 18 | | | | | | | | |
| 19 | | | | | | | | |

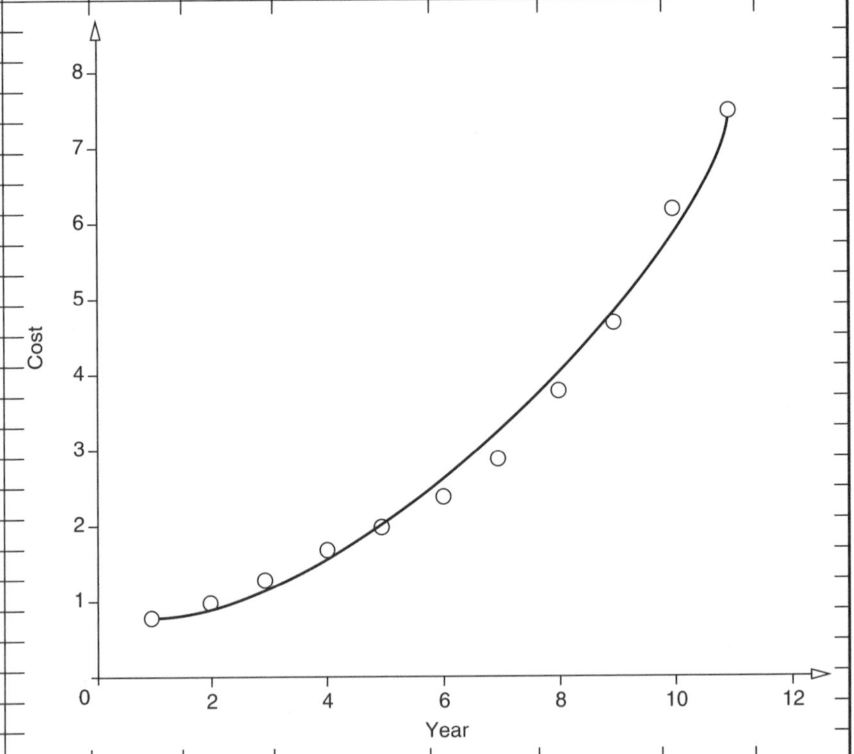

**Figure 7.12** Non-linear regression for Worked Example 7.10.

The spreadsheet has calculated the line of best fit as:

$$y = bm^x = 0.6541 \times 1.2475^x$$

Then it used this equation to find the trend in prices charged by the accountant over the past eleven years, and plotted the results. The value of 1.2475 for $m$ suggests that the accountant's charges are rising by almost 25% a year, which is certainly more than inflation.

## In summary

**Non-linear regression finds the equation for the best curve through a set of data. The calculations for this can become quite complicated.**

# Chapter review

This chapter described the relationship between variables. It concentrated on linear regression, which finds the line of best fit through a set of data. In particular, it:

- discussed errors and noise;
- used the mean squared error to measure the noise;
- used linear regression to find the line of best fit;
- discussed the use of linear regression for forecasting;
- described the coefficients of determination and correlation;
- discussed multiple linear regression;
- mentioned non-linear regression.

# Problems

7.1. The productivity of a factory has been recorded over a ten-month period, together with forecasts made by the production manager, foreman and management services department shown below. How accurate are these forecasts?

**Table 7.4**

| Month | 1 | 2 | 3 | 4 | 5 | 6 | 7 | 8 | 9 | 10 |
|-------|---|---|---|---|---|---|---|---|---|----|
| Productivity | 22 | 24 | 28 | 27 | 23 | 24 | 20 | 18 | 20 | 23 |
| Production manager | 23 | 26 | 32 | 28 | 20 | 26 | 24 | 16 | 21 | 23 |
| Foreman | 22 | 28 | 29 | 29 | 24 | 26 | 21 | 21 | 24 | 25 |
| Management services | 21 | 25 | 26 | 27 | 24 | 23 | 20 | 20 | 19 | 24 |

7.2 Find the line of best fit through the following data. How good is this fit?

| x | 10 | 19 | 29 | 42 | 51 | 60 | 73 | 79 | 90 | 101 |
|---|----|----|----|----|----|----|----|----|----|-----|
| y | 69 | 114 | 163 | 231 | 272 | 299 | 361 | 411 | 483 | 522 |

7.3 A local amateur dramatics society is staging a play and wants to know how much to spend on advertising. The spending on advertising (in hundreds of pounds) and audience size for the past eleven productions is shown in the following table. If the hall capacity is now 300 people how much should it spend on advertising?

| Spending | 3 | 5 | 1 | 7 | 2 | 4 | 4 | 2 | 6 | 6 | 4 |
|----------|---|---|---|---|---|---|---|---|---|---|---|
| Audience | 200 | 250 | 75 | 425 | 125 | 300 | 225 | 200 | 300 | 400 | 275 |

7.4 Ten experiments were carried out to check the effects of bonus rates paid to salespeople on sales, with the following results.

| % Bonus | 0 | 1 | 2 | 3 | 4 | 5 | 6 | 7 | 8 | 9 |
|---------|---|---|---|---|---|---|---|---|---|---|
| Sales ('00) | 3 | 4 | 8 | 10 | 15 | 18 | 20 | 22 | 27 | 28 |

What is the line of best fit through these data? How good is this line?

7.5 Sales of a product for the past 10 months are shown below. Use linear regression to forecast sales for the next 6 months. How reliable are these figures?

| Month | 1 | 2 | 3 | 4 | 5 | 6 | 7 | 8 | 9 | 10 |
|-------|---|---|---|---|---|---|---|---|---|-----|
| Sales | 6 | 21 | 41 | 75 | 98 | 132 | 153 | 189 | 211 | 243 |

7.6 A company records sales of four products for a ten-month period shown below. What can you say about these figures?

| Month | 1 | 2 | 3 | 4 | 5 | 6 | 7 | 8 | 9 | 10 |
|-------|---|---|---|---|---|---|---|---|---|-----|
| p | 24 | 36 | 45 | 52 | 61 | 72 | 80 | 94 | 105 | 110 |
| q | 2500 | 2437 | 2301 | 2290 | 2101 | 2001 | 1995 | 1847 | 1732 | 1695 |
| r | 150 | 204 | 167 | 254 | 167 | 241 | 203 | 224 | 167 | 219 |
| s | 102 | 168 | 205 | 221 | 301 | 302 | 310 | 459 | 519 | 527 |

# CS Case study – Western General Hospital

Each term the Western General Hospital accepts a batch of 50 new student nurses. Their training lasts for several years before they become state registered or state enrolled. Unfortunately, many nurses fail exams and do not complete their training. If the hospital can reduce the number of nurses leaving they can save a considerable amount of the cost of training.

One administrator has suggested that the recruitment procedure is at fault, and they should take more care to choose students who are likely to complete the course. One way of doing this could be to relate the nurses' likely performance in exams to their performance in school exams. But nurses come from a variety of backgrounds and start training at different ages, so their performance at school may not be relevant. Other possible factors are age and number of previous jobs.

Table 7.6 shows results for last term's nurses. Exam grades have been converted to numbers (A = 5, B = 4 and so on), and average marks are given.

**Table 7.6**

| Nurse | Year of birth | Nursing grade | School grade | Number of jobs | Nurse | Year of birth | Nursing grade | School grade | Number of jobs |
|---|---|---|---|---|---|---|---|---|---|
| 1 | 72 | 2.3 | 3.2 | 0 | 26 | 60 | 4.1 | 3.7 | 4 |
| 2 | 65 | 3.2 | 4.5 | 1 | 27 | 74 | 2.6 | 2.3 | 1 |
| 3 | 72 | 2.8 | 2.1 | 1 | 28 | 74 | 2.3 | 2.7 | 1 |
| 4 | 62 | 4.1 | 1.6 | 4 | 29 | 72 | 1.8 | 1.9 | 2 |
| 5 | 70 | 4.0 | 3.7 | 2 | 30 | 71 | 3.1 | 1.0 | 0 |
| 6 | 73 | 3.7 | 2.0 | 1 | 31 | 62 | 4.8 | 1.2 | 3 |
| 7 | 65 | 3.5 | 1.5 | 0 | 32 | 68 | 2.3 | 3.0 | 1 |
| 8 | 63 | 4.8 | 3.6 | 0 | 33 | 70 | 3.1 | 2.1 | 5 |
| 9 | 73 | 2.8 | 3.4 | 2 | 34 | 71 | 2.2 | 4.0 | 2 |
| 10 | 74 | 1.9 | 1.2 | 1 | 35 | 72 | 3.0 | 4.5 | 3 |
| 11 | 74 | 2.3 | 4.8 | 2 | 36 | 62 | 4.3 | 3.3 | 0 |
| 12 | 73 | 2.5 | 4.5 | 0 | 37 | 72 | 2.4 | 3.1 | 1 |
| 13 | 66 | 2.8 | 1.0 | 0 | 38 | 68 | 3.2 | 2.9 | 0 |
| 14 | 59 | 4.5 | 2.2 | 3 | 39 | 74 | 1.1 | 2.5 | 0 |
| 15 | 74 | 2.0 | 3.0 | 1 | 40 | 59 | 4.2 | 1.9 | 2 |
| 16 | 70 | 3.4 | 4.0 | 0 | 41 | 68 | 2.0 | 1.2 | 1 |
| 17 | 68 | 3.0 | 3.9 | 2 | 42 | 74 | 1.0 | 4.1 | 0 |
| 18 | 68 | 2.5 | 2.9 | 2 | 43 | 67 | 3.0 | 3.0 | 0 |
| 19 | 69 | 2.8 | 2.0 | 1 | 44 | 70 | 2.0 | 2.2 | 0 |
| 20 | 71 | 2.8 | 2.1 | 1 | 45 | 66 | 2.3 | 2.0 | 2 |
| 21 | 68 | 2.7 | 3.8 | 0 | 46 | 66 | 3.7 | 3.7 | 4 |
| 22 | 61 | 4.5 | 1.4 | 3 | 47 | 58 | 4.7 | 4.0 | 5 |
| 23 | 65 | 3.7 | 1.8 | 2 | 48 | 65 | 4.0 | 1.9 | 2 |
| 24 | 70 | 3.0 | 2.4 | 6 | 49 | 65 | 3.8 | 3.1 | 0 |
| 25 | 71 | 2.9 | 3.0 | 0 | 50 | 69 | 2.5 | 4.6 | 1 |

When the hospital collected data on the number of nurses who did not finish training in the past ten terms, they got the following results:

| Term | 1 | 2 | 3 | 4 | 5 | 6 | 7 | 8 | 9 | 10 |
|--------|---|---|---|---|---|----|----|----|----|----|
| Number | 4 | 7 | 3 | 6 | 9 | 11 | 10 | 15 | 13 | 17 |

What can the hospital do with these figures?

# 8 Business forecasting

**Contents**

## Chapter outline

This chapter looks at different ways of forecasting. After reading the chapter you should be able to:

- understand the importance of forecasting in organisations
- list different ways of forecasting
- describe judgemental forecasting
- forecast using actual averages, moving averages and exponential smoothing
- make forecasts for time series with seasonality and trend

# 8.1 Different ways of forecasting

All decisions become effective at some point in the future. When you decide to go to the cinema, or build a new factory, or invest in gold, you make a decision at one point and it becomes effective some time later. Thus your decision is not based on present circumstances, but on circumstances in the future when the decision becomes effective. So all decisions are based on forecasts of future circumstances.

There are many ways of making forecasts. Because of the wide range of things to be forecast, and the different circumstances, there is no single best method. Nevertheless, we can describe some general methods of forecasting, and say when each can be used.

We can classify forecasting methods in several ways. We can, for example, talk about the time in the future covered by forecasts.

- **Long-term forecasts** look ahead several years – perhaps the time needed to build a new factory.

- **Medium-term forecasts** look ahead between three months and two years – the time typically needed to replace an old product by a new one.

- **Short-term forecasts** cover the next few weeks – describing the continuing demand for a product.

Figure 8.1 shows another way of looking at forecasting. If a company already makes a product, it will have some records and can use a quantitative method to forecast future demand. There are two ways of doing this:

- **Projective methods** which examine the pattern of past demand and extend this pattern into the future. If demand in the past four weeks has been 10, 20, 30 and 40, it is reasonable to project this pattern and to say that demand in the following week will be around 50.

- **Causal methods** which analyse the effects of outside influences and use these to make forecasts. The productivity of a factory might depend on the bonus rates paid to employees – so we can use the proposed bonus rate to forecast future productivity. We have already met this approach with linear regression in Chapter 7.

Both of these approaches need numerical data about past performance. However if, for example, a company starts making an entirely new product, it will not have this historical data. When there is no quantitative data we have to use a qualitative forecasting method.

In Chapter 7 we looked at linear regression for making causal forecasts, so the rest of this chapter will describe the other methods.

## In summary

**Forecasting is needed for all decisions. There is no single, best method so here we will describe several approaches. We have already looked at causal forecasting, so will concentrate on judgemental and projective methods.**

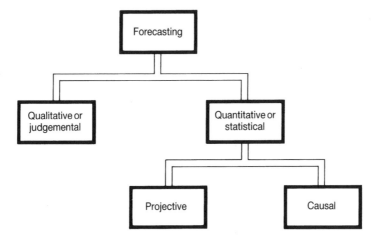

**Figure 8.1**  Classification of forecasting methods.

# 8.2  Judgemental forecasting

Suppose a company is about to market an entirely new product, or a medical team is trying a new organ transplant, or a board of directors is considering plans for 25 years in the future. None of these groups has any historical data for a quantitative forecast. Sometimes there simply are no data, and sometimes the available data are unreliable or irrelevant to the future.

When organisations cannot use a quantitative forecast, they have to use a qualitative one. These are based on expert opinions. Five widely used methods are:

- personal insight
- panel consensus
- market surveys
- historical analogy
- Delphi method

## 8.2.1  Personal insight

Personal insight involves a single expert who is familiar with the situation making a forecast based on his or her own judgement. This is the most widely used forecasting method – and is the one that managers should try to avoid. It relies entirely on one person's judgement – and their opinions, prejudices and ignorance. It can give good forecasts but often gives very bad ones.

The major weakness of this method is its unreliability. Comparisons have shown clearly that someone who is familiar with a situation, but uses experience and subjective opinions to forecast will consistently make worse forecasts than someone who knows nothing about the situation but uses a more formal method.

## 8.2.2 Panel consensus

A single expert can easily make a mistake, but collecting several experts and encouraging them to talk freely should lead to a consensus that is more reliable. This is certainly better than one person's insight, but it can be difficult to combine the views of different people when they cannot reach a consensus; and even groups of experts can make mistakes. There are also the problems of group working, where 'he who shouts loudest gets his way', everyone tries to please the boss, some people do not speak well in groups, and so on.

## 8.2.3 Market surveys

Sometimes, even groups of experts do not know enough about a situation to make a forecast. This happens with the launch of a new product, when it is better to talk directly to potential customers. Market surveys collect data from a representative sample of customers. The customers' views are analysed, and inferences are drawn about the whole population, as described in Chapter 3.

   The main drawbacks with market surveys are that they are expensive and time consuming, and they rely on:

- a sample of customers that fairly represents the population;

- useful, unbiased questions;

- reliable analyses of the replies;

- valid conclusions drawn from the analyses.

## 8.2.4 Historical analogy

If a company is introducing a new product, it may have a similar product that was launched recently. Then the demand for the new product might follow the same pattern. If, for example, a publisher is introducing a new book, it can forecast likely demand from the actual demand for a similar book that it published recently.

## 8.2.5 Delphi method

A number of experts are contacted by post and each is given a questionnaire to complete. The replies from these questionnaires are analysed and summaries are passed back to the experts. Each expert is then asked to reconsider their original reply in the light of the summarised replies from others. This process of modifying responses in the light of replies made by the rest of the group is repeated several times – usually between three and six. By this time, the range of opinions should have narrowed enough to help with decisions.

## In summary

**Judgemental forecasts are used when there are no relevant historical data. They rely on the subjective views of experts. Common methods are personal insight, panel consensus, market surveys, historical analogy and the Delphi method.**

# 8.3  Projective forecasting

## 8.3.1  Time series

Projective forecasting uses historical observations to forecast future values. Many of these observations occur as **time series**, which are series of observations taken at regular intervals of time – like monthly unemployment figures, daily rainfall, weekly demand and annual population statistics.

If you have a time series, a scatter diagram will often show the underlying patterns. The three most common patterns are:

- **Constant series**, where observations take roughly the same value over time, like annual rainfall.

- **Series with a trend**, which either rise or fall steadily, like the gross national product per capita.

- **Seasonal series**, which follow cycles, like the weekly sales of suntan lotion.

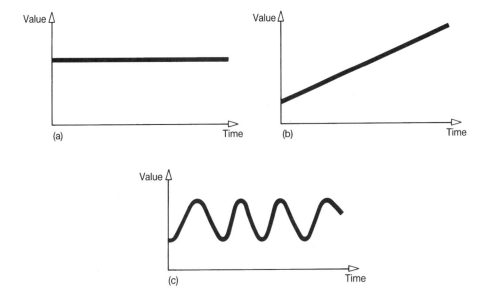

**Figure 8.2**   Common patterns in time series: (a) constant series; (b) series with a trend; (c) seasonal series.

If observations followed these simple patterns we would have no problems with forecasting. But as we saw with regression, there is nearly always some random noise that creates a difference between the underlying pattern and the actual observations. A constant series, for example, does not always take exactly the same value, but is somewhere close, like:

$$200 \quad 205 \quad 194 \quad 195 \quad 208 \quad 203 \quad 200 \quad 193 \quad 201 \quad 198$$

Figure 8.3 illustrates the effect of noise on the standard patterns. While the noise remains relatively small we can get good forecasts; but if there is a lot of noise then the underlying pattern becomes hidden and forecasting becomes more difficult.

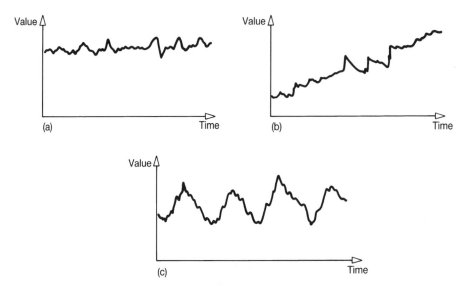

**Figure 8.3** Common patterns in time series including noise: (a) constant series; (b) series with a trend; (c) seasonal series.

As before, we can define an error in every observation.

$E_t$ = actual observation in period $t$ – forecast value
      = $y_t - F_t$

where:

$y_t$ = actual observation in period $t$

$F_t$ = value forecast in period $t$ from the underlying pattern

We can again use the mean squared error to summarise these individual results.

$$\text{mean squared error} = \frac{\Sigma(E_t)^2}{n} = \frac{\Sigma(y_t - F_t)^2}{n}$$

# we Worked example 8.1

Two forecasting methods give the results shown in Table 8.1 for a time series. Which method is more accurate?

**Table 8.1**

| $t$ | 1 | 2 | 3 | 4 | 5 |
|---|---|---|---|---|---|
| $y_t$ | 20 | 22 | 26 | 19 | 14 |
| $F_t$ with method 1 | 17 | 23 | 24 | 22 | 17 |
| $F_t$ with method 2 | 21 | 20 | 228 | 24 | 19 |

## Solution

Figure 8.4 shows a spreadsheet calculating the errors. The mean squared errors are 6.4 and 19.0 respectively, so method 1 gives the better forecasts.

| | A | B | C | D | E | F | G | H |
|---|---|---|---|---|---|---|---|---|
| 1 | **Forecasting errors** | | | | | | | |
| 2 | | | | | | | | |
| 3 | | | | | | | | |
| 4 | **Period** | **Value** | **Forecast 1** | **Error** | **Error squared** | **Forecast 2** | **Error** | **Error squared** |
| 5 | 1 | 20 | 17 | 3 | 9 | 15 | 5 | 25 |
| 6 | 2 | 22 | 23 | −1 | 1 | 20 | 2 | 4 |
| 7 | 3 | 26 | 24 | 2 | 4 | 22 | 4 | 16 |
| 8 | 4 | 19 | 22 | −3 | 9 | 24 | −5 | 25 |
| 9 | 5 | 14 | 17 | −3 | 9 | 19 | −5 | 25 |
| 10 | **Totals** | 101 | 103 | −2 | 32 | 100 | 1 | 95 |
| 11 | **Means** | 20.2 | 20.6 | −0.4 | 6.4 | 20 | 0.2 | 19 |

**Figure 8.4** Calculating the mean squared errors for Worked Example 8.1.

## In summary

Projective forecasts use historical observations to forecast future values. Projective forecasts often use time series. Observations usually follow an underlying pattern with superimposed noise.

## 8.3.2 Actual averages

Suppose you are booking a holiday to Florida and want to know the likely temperature during your stay. The easiest way of finding this is to look up figures for past years and take an average. If your holiday starts on 1 July you can find the average temperature on 1 July over, say, the past twenty years. This is an example of forecasting using actual averages.

**Actual averages forecast**   $F_{t+1} = \dfrac{\Sigma y_t}{n}$

where:

$n$ = number of periods of historical data

$y_t$ = observation at time $t$

$F_{t+1}$ = forecast for time $t+1$

## we Worked example 8.2

Use actual averages to forecast values for period 6 of the following two time series. How accurate are the forecasts? What are the forecasts for period 24?

| Period | 1 | 2 | 3 | 4 | 5 |
|--------|-----|-----|-----|-----|-----|
| Series 1 | 98 | 100 | 98 | 104 | 100 |
| Series 2 | 140 | 66 | 152 | 58 | 84 |

## Solution

Series 1:  $F_6 = \dfrac{\Sigma y_t}{n} = \dfrac{1}{5} \times 500 = 100$

Series 2:  $F_6 = \dfrac{500}{5} = 100$

Actual averages give the same forecasts for period 6. But there is obviously less noise in the first series than in the second, so we would expect this result to be more accurate.

Actual averages assume that the underlying pattern is constant. So the forecasts for period 24 are the same as the forecasts for period for 6 (i.e. 100).

Using actual averages to forecast is easy and can work well for constant values, but the method does not work well if the pattern changes. Older data tend to swamp the latest figures and the forecast is unresponsive to the change.

Suppose that demand for an item has been constant at 100 units a week for the past two years. Actual averages give a forecast demand for week 105 of 100 units. If the actual demand in week 105 suddenly rises to 200 units, actual averages give a forecast for week 106 of:

$$F_{106} = \frac{(104 \times 100) + 200}{105} = 100.95$$

A rise in demand of 100 gives an increase of 0.95 in the forecast. If demand continues at 200 units a week, the subsequent forecasts are:

$$F_{107} = 101.89 \quad F_{108} = 102.80 \quad F_{109} = 103.70, \text{ and so on}$$

The forecasts are rising but the response is slow.

Because very few time series are stable over long periods, actual averages are not widely used.

## In summary

**Actual averages give good results with a constant underlying value but they do not respond well to changes. Hence we usually have to use other methods.**

## 8.3.3 Moving averages

If circumstances change over time, only a certain amount of historical data is relevant for future decisions. So we might use the average value over, say, the past six weeks as a forecast, and ignore any data that is older than this. This gives a **moving average**. We always take the average of the latest $n$ periods of data, and as new data become available we discard the older data.

$$
\begin{aligned}
F_{t+1} &= \text{average of n most recent pieces of data} \\
&= \frac{(\text{latest demand} + \text{next latest} + \ldots + n_{\text{th}} \text{ latest})}{n} \\
&= \frac{(y_t + y_{t-1} + \ldots\ldots y_{t-n+1})}{n}
\end{aligned}
$$

## we Worked example 8.3

In the past few months, Consolidated Trucking has dealt with the following numbers of customer enquiries.

| $t$ | 1 | 2 | 3 | 4 | 5 | 6 | 7 | 8 |
|-----|-----|-----|-----|-----|-----|-----|-----|-----|
| $y_t$ | 135 | 130 | 125 | 135 | 115 | 80 | 105 | 100 |

Changing conditions within the company mean that any data over three months old is irrelevant. How would you forecast the number of enquiries in the future?

## Solution

Only data more recent than three months are relevant, so we can use a three-month moving average for the forecast. At the end of period 3, the forecast for period 4 is:

$$F_4 = \frac{y_1 + y_2 + y_3}{3} = \frac{135 + 130 + 125}{3} = 130$$

At the end of period 4, we update this forecast to give:

$$F_5 = \frac{y_2 + y_3 + y_4}{3} = \frac{130 + 125 + 135}{3} = 130$$

Then:

$$F_6 = \frac{y_3 + y_4 + y_5}{3} = \frac{125 + 135 + 115}{3} = 125$$

$$F_7 = \frac{y_4 + y_5 + y_6}{3} = \frac{135 + 115 + 80}{3} = 110$$

$$F_8 = \frac{y_5 + y_6 + y_7}{3} = \frac{115 + 80 + 105}{3} = 100$$

$$F_9 = \frac{y_6 + y_7 + y_8}{3} = \frac{80 + 105 + 100}{3} = 95$$

In Worked Example 8.3, you can see clearly that the forecast is responding to changes. A high value in one month moves the forecast upwards, and a low number moves the forecast downwards. The rate at which a forecast responds to changing demand gives its **sensitivity**. We can adjust the sensitivity of a moving average by changing the value of $n$. A high value of $n$ takes the average of a large number of observations, making the forecast insensitive: it smooths out random variations, but is slow to follow real changes. A small value of $n$ gives a sensitive forecast that quickly follows real changes, but it may be too sensitive to random fluctuations. We need a compromise value of $n$ which gives reasonable results – this is typically around six periods.

# we Worked example 8.4

The following table shows monthly demand for a product over the past year. Use moving averages with $n = 3$, $n = 6$ and $n = 9$ to get forecasts for one period ahead in each instance.

| Month | 1 | 2 | 3 | 4 | 5 | 6 | 7 | 8 | 9 | 10 | 11 | 12 |
|---|---|---|---|---|---|---|---|---|---|---|---|---|
| Demand | 16 | 14 | 12 | 15 | 18 | 21 | 23 | 24 | 25 | 26 | 37 | 38 |

## Solution

The earliest forecast we can make using a three-period moving average is

$$F_4 = (y_1 + y_2 + y_3)/3$$

Similarly, the earliest forecast for a six- and nine-period moving average are $F_7$ and $F_{10}$ respectively. Figure 8.5 shows the forecasts in a spreadsheet.

| | A | B | C | D | E |
|---|---|---|---|---|---|
| 1 | **Moving average forecasts** | | | | |
| 2 | | | | | |
| 3 | | | **Forecasts** | | |
| 4 | **Month** | **Demand** | **n = 3** | **n = 6** | **n = 9** |
| 5 | 1 | 16 | | | |
| 6 | 2 | 14 | | | |
| 7 | 3 | 12 | | | |
| 8 | 4 | 15 | 14.00 | | |
| 9 | 5 | 18 | 13.67 | | |
| 10 | 6 | 21 | 15.00 | | |
| 11 | 7 | 23 | 18.00 | 16.00 | |
| 12 | 8 | 24 | 20.67 | 17.17 | |
| 13 | 9 | 25 | 22.67 | 18.83 | |
| 14 | 10 | 26 | 24.00 | 21.00 | 18.67 |
| 15 | 11 | 37 | 25.00 | 22.83 | 19.78 |
| 16 | 12 | 38 | 29.33 | 26.00 | 22.33 |
| 17 | 13 | | 33.67 | 28.83 | 25.22 |

**Figure 8.5**  Moving average forecasts for Worked Example 8.4.

Figure 8.6 shows a graph of these results, where the demand is clearly rising. The three-month moving average is following this trend quite quickly, while the nine-month moving average is least responsive.

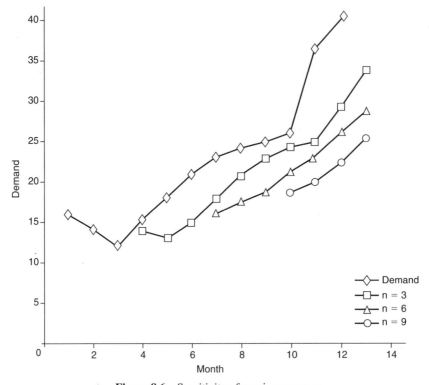

**Figure 8.6** Sensitivity of moving averages.

Although moving averages are more responsive than actual averages, they still have a number of defects, including:

● they give the same weight to all observations;

● they only work well with constant time series;

● they need a lot of historical data to update forecast;

● the choice of $n$ is often arbitrary.

We can overcome most of these problems using exponential smoothing, which is described in the following section.

## In summary

**Moving averages give forecasts based on the latest $n$ observations and ignore all older values. We can change the sensitivity by altering the value of $n$.**

### 8.3.4 Exponential smoothing

Exponential smoothing is currently the most widely used forecasting method. It is based on the idea that as data get older they become less relevant and should be given less weight. In particular, exponential smoothing gives a declining weight to observations, as shown in Figure 8.7.

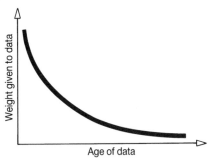

**Figure 8.7**   Weight given to data with exponential smoothing.

We can get this declining weight using only the latest observation and the previous forecast.

> **Exponential smoothing**
> new forecast = {$\alpha$ × latest observation} + {$(1 - \alpha)$ × last forecast}
> or        $F_{t+1} = \alpha y_t + (1 - \alpha)F_t$

In this equation, $\alpha$ is the smoothing constant which usually takes a value between 0.1 and 0.2. You can see how exponential smoothing adapts to changes with a simple example.

Suppose a forecast was optimistic and suggested a value of 200 ($= F_t$) for a demand that actually turns out to be 180 ($= y_t$). Taking a value of $\alpha = 0.2$, the forecast for the next period is:

$$F_{t+1} = \alpha y_t + (1 - \alpha)F_t = \{0.2 \times 180\} + \{(1 - 0.2) \times 200\} = 196$$

The optimistic forecast is noted and the value for the next period is adjusted downwards. You can see the reason for this adjustment if we rearrange the exponential smoothing formula:

$$F_{t+1} = \alpha y_t + (1 - \alpha)F_t = F_t + \alpha(y_t - F_t)$$

but we said earlier that the error, $E_t$, is:

$$E_t = y_t - F_t$$

so

$$F_{t+1} = F_t + \alpha E_t$$

The error in each forecast is noted, and a proportion is added to adjust the next forecast. The larger the error in the last forecast, the greater is the adjustment for the next forecast.

# we Worked example 8.5

Use exponential smoothing with $\alpha = 0.2$ and an initial value of $F_1 = 170$ to get forecasts for one period ahead for the following time series.

| Month | 1 | 2 | 3 | 4 | 5 | 6 | 7 | 8 |
|-------|-----|-----|-----|-----|-----|-----|-----|-----|
| Demand | 178 | 180 | 156 | 150 | 162 | 158 | 154 | 132 |

## Solution

We know that $F_1 = 170$ and $\alpha = 0.2$. Substituting these values gives:

$$F_2 = \alpha y_1 + (1-\alpha)F_1 = (0.2 \times 178) + (0.8 \times 170) = 171.6$$

$$F_3 = \alpha y_2 + (1-\alpha)F_2 = (0.2 \times 180) + (0.8 \times 171.6) = 173.3$$

$$F_4 = \alpha y_3 + (1-\alpha)F_3 = (0.2 \times 156) + (0.8 \times 173.3) = 169.8$$

and so on, as shown in Figure 8.8.

|   | A | B | C |
|---|---|---|---|
| 1 | **Exponential smoothing** | | |
| 2 | | | |
| 3 | **Month** | **Demand** | **Forecast** |
| 4 | 1 | 178 | 170.00 |
| 5 | 2 | 180 | 171.60 |
| 6 | 3 | 156 | 173.28 |
| 7 | 4 | 150 | 169.82 |
| 8 | 5 | 162 | 165.86 |
| 9 | 6 | 158 | 165.09 |
| 10 | 7 | 154 | 163.67 |
| 11 | 8 | 132 | 161.74 |
| 12 | 9 | | 155.79 |

**Figure 8.8**   Exponential smoothing forecasts for Worked Example 8.5.

The value given to the smoothing constant, $\alpha$, is important in setting the sensitivity of the forecasts as it dictates the balance between the last forecast and the latest observation. So a high value of $\alpha$ (say between 0.3 and 0.35) gives a sensitive forecast; a lower value of $\alpha$ (say between 0.05 and 0.1) gives a less sensitive forecast. Again we need a compromise between a sensitive forecast that might follow random fluctuations and an insensitive one that might miss real patterns.

## we Worked Example 8.6

The following time series has a clear step upwards in demand in month 3. Use an initial forecast of 500 to compare exponential smoothing forecasts with varying values of $\alpha$.

| Month | 1 | 2 | 3 | 4 | 5 | 6 | 7 | 8 | 9 | 10 | 11 |
|---|---|---|---|---|---|---|---|---|---|---|---|
| Demand | 480 | 500 | 1500 | 1450 | 1550 | 1500 | 1480 | 1520 | 1500 | 1490 | 1500 |

## Solution

Taking values of $\alpha=0.1$, 0.2, 0.3 and 0.4 gives the results shown in Figure 8.9. All the forecasts will eventually follow the sharp step and raise forecasts to around 1500. Higher values of $\alpha$ make this adjustment more quickly, as shown in Figure 8.10.

| | A | B | C | D | E | F |
|---|---|---|---|---|---|---|
| | | | | | | |
| 1 | **Exponential smoothing** | | | | | |
| 2 | | | | | | |
| 3 | | | **Forecasts** | | | |
| 4 | **Month** | **Demand** | **$\alpha = 0.1$** | **$\alpha = 0.2$** | **$\alpha = 0.3$** | **$\alpha = 0.4$** |
| 5 | 1 | 480 | 500.00 | 500.00 | 500.00 | 500.00 |
| 6 | 2 | 500 | 498.00 | 496.00 | 494.00 | 492.00 |
| 7 | 3 | 1500 | 498.20 | 496.80 | 495.80 | 495.20 |
| 8 | 4 | 1450 | 598.38 | 697.44 | 797.06 | 897.12 |
| 9 | 5 | 1550 | 683.54 | 847.95 | 992.94 | 1118.27 |
| 10 | 6 | 1500 | 770.19 | 988.36 | 1160.06 | 1290.96 |
| 11 | 7 | 1480 | 843.17 | 1090.69 | 1262.04 | 1374.58 |
| 12 | 8 | 1520 | 906.85 | 1168.55 | 1327.43 | 1416.75 |
| 13 | 9 | 1500 | 968.17 | 1238.84 | 1385.20 | 1458.05 |
| 14 | 10 | 1490 | 1021.35 | 1291.07 | 1419.64 | 1474.83 |
| 15 | 11 | 1500 | 1068.22 | 1330.86 | 1440.75 | 1480.90 |
| 16 | 12 | | 1111.39 | 1364.69 | 1458.52 | 1488.54 |

**Figure 8.9** Calculations for exponential smoothing in Worked Example 8.6.

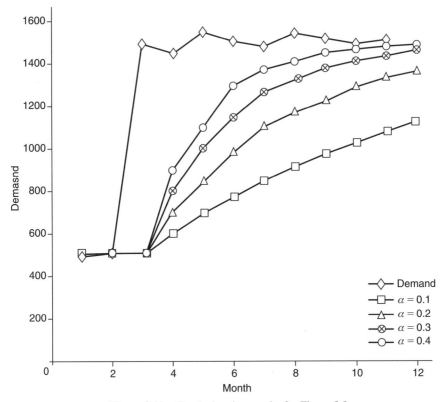

**Figure 8.10**  Graph showing results for Figure 8.9.

Although higher values of $\alpha$ give more responsive forecasts, they do not necessarily give more accurate ones. Observations always contain noise, and very sensitive forecasts tend to follow these random fluctuations. The usual way to choose $\alpha$ is to test several values over a trial period and then use the one that gives smallest errors.

## In summary

**Exponential smoothing adds portions of the last forecast and the latest observation. This reduces the weight given to data as its age increases. The smoothing constant sets the sensitivity of the forecast.**

# 8.4 Forecasting with seasonality and trend

The methods described so far give good results for constant time series, but they do not work well with other patterns. There are several ways of forecasting more complex time series, but the easiest is to split observations into different components and to forecast each component separately. Then we get the final forecast by combining the separate components.

Typically, observations are made-up of three components:

- **Trend value** $(T)$ – an underlying upward or downward movement over time.
- **Seasonal factor** $(S)$ – regular variation around the trend, typically showing variations over a year.
- **Error** $(E)$ – random noise.

So, to forecast we:

- deseasonalise the data and find the trend value, $T$;
- find the seasonal indices, $S$;
- use the calculated trend and seasonal indices to forecast $F = T \times S$.

The easiest way of finding the trend value is to use linear regression with time as the independent variable.

## we Worked example 8.7

Use linear regression to find the deseasonalised trend for the following observations.

| Period | 1 | 2 | 3 | 4 | 5 | 6 | 7 | 8 | 9 | 10 | 11 | 12 |
|---|---|---|---|---|---|---|---|---|---|---|---|---|
| Observation | 291 | 320 | 142 | 198 | 389 | 412 | 271 | 305 | 492 | 518 | 363 | 388 |

## Solution

With the values given:

$$n = 12 \quad \Sigma x = 78 \quad \Sigma y = 4089 \quad \Sigma x^2 = 650 \quad \Sigma(xy) = 29\,160$$

Substituting these in the standard linear regression equations gives:

$$b = \frac{n\Sigma(xy) - \Sigma x \Sigma y}{n\Sigma x^2 - (\Sigma x)^2} = \frac{(12 \times 29\,160) - (78 \times 4089)}{(12 \times 650) - (78 \times 78)} = 18.05$$

$$a = \bar{y} - b\bar{x} = \frac{4089}{12} - 18.05 \times \frac{78}{12} = 223.41$$

Figure 8.11 shows the calculations in a spreadsheet, and this confirms the line of best fit as:

observation $= 223.41 + (18.05 \times$ period)

So the deseasonalised trend value for period 1 is:

$223.41 + (18.05 \times 1) = 241.46$

Substituting other values for the period gives the deseasonalised trend line shown in Figure 8.12.

| | A | B | C | D | E |
|---|---|---|---|---|---|
| 1 | **Seasonal trend** | | | | |
| 2 | | | | | |
| 3 | **Period** | **Observation** | | **Deseasonalised** | |
| 4 | | | | **trend value** | |
| 5 | 1 | 291 | | 241.46 | |
| 6 | 2 | 320 | | 259.51 | |
| 7 | 3 | 142 | | 277.57 | |
| 8 | 4 | 198 | | 295.62 | |
| 9 | 5 | 389 | | 313.67 | |
| 10 | 6 | 412 | | 331.72 | |
| 11 | 7 | 271 | | 349.78 | |
| 12 | 8 | 305 | | 367.83 | |
| 13 | 9 | 492 | | 385.88 | |
| 14 | 10 | 518 | | 403.93 | |
| 15 | 11 | 363 | | 421.99 | |
| 16 | 12 | 388 | | 440.04 | |
| 17 | | | | | |
| 18 | **Summary results** | | | | |
| 19 | | | | | |
| 20 | *Regression statistics* | | | | |
| 21 | Correlation, r | 0.591 | | | |
| 22 | Determination r^2 | 0.349 | | | |
| 23 | Adjusted r^2 | 0.284 | | | |
| 24 | Observations | 12 | | | |
| 25 | | | | | |
| 26 | *Coefficients* | | | | |
| 27 | Intercept | 223.409 | | | |
| 28 | X variable 1 | 18.052 | | | |

**Figure 8.11**   Using linear regression to deseasonalise values for Worked Example 8.7.

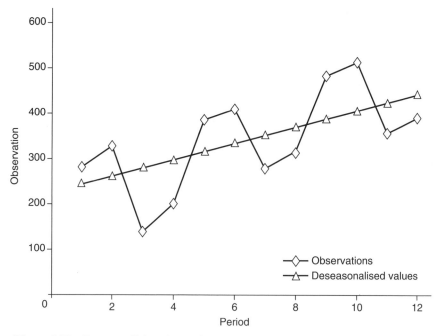

**Figure 8.12** Deseasonalising observations with regression line for Worked Example 8.7.

Now we have the deseasonalised trend value, so the next step is to find the seasonal indices. The seasonal index is the amount we multiply the deseasonalised value by to get the seasonal value.

$$\text{seasonal index} = \frac{\text{seasonal value}}{\text{deseasonalised value}}$$

Suppose a newspaper has average daily sales of 1000 copies, but this rises to 2000 copies on Saturday and falls to 500 copies on Monday and Tuesday. The deseasonalised value is 1000, the seasonal index for Saturday is 2000/1000 = 2.0, the seasonal indices for Monday and Tuesday are 500/1000 = 0.5 and seasonal indices for other days are 1000/1000 = 1.0.

We can take the actual observation for a period and divide by the trend value to get a seasonal index for the period. This index is affected by noise in the data, so it is only an approximation. However, if we take several complete seasons we can find average indices that are more reliable.

# we Worked example 8.8

Find the seasonal indices for the data in Worked Example 8.7.

## Solution

Taking a single period, say 4, we have an actual observation of 198. The deseasonalised value using linear regression is 295.61, so the seasonal index is:

$$\frac{198}{295.61} = 0.67.$$

Figure 8.13 shows the results of repeating these calculations for other periods.

|   | A | B | C | D | E | F |
|---|---|---|---|---|---|---|
| 1 | **Seasonal indices** | | | | | |
| 2 | | | | | | |
| 3 | **Period** | **Observation** | **Deseasonalised** | **Seasonal** | **Period in** | **Average** |
| 4 | | | **trend value** | **index** | **season** | **index** |
| 5 | 1 | 291 | 241.46 | 1.21 | 1 | 1.24 |
| 6 | 2 | 320 | 259.51 | 1.23 | 2 | 1.25 |
| 7 | 3 | 142 | 277.57 | 0.51 | 3 | 0.71 |
| 8 | 4 | 198 | 295.62 | 0.67 | 4 | 0.79 |
| 9 | 5 | 389 | 313.67 | 1.24 | 1 | 1.24 |
| 10 | 6 | 412 | 331.72 | 1.24 | 2 | 1.25 |
| 11 | 7 | 271 | 349.78 | 0.77 | 3 | 0.71 |
| 12 | 8 | 305 | 367.83 | 0.83 | 4 | 0.79 |
| 13 | 9 | 492 | 385.88 | 1.28 | 1 | 1.24 |
| 14 | 10 | 518 | 403.93 | 1.28 | 2 | 1.25 |
| 15 | 11 | 363 | 421.99 | 0.86 | 3 | 0.71 |
| 16 | 12 | 388 | 440.04 | 0.88 | 4 | 0.79 |

**Figure 8.13**   Finding the seasonal indices.

We know from Figure 8.12 that there are four periods in a season – in other words, there is a basic pattern in the data that repeats every four periods. So we need to find four seasonal indices.

Periods 1, 5 and 9 are the first periods in consecutive seasons. Taking the average of these indices gives a better value for the index as:

● Average seasonal index for first period in season $(1.21 + 1.24 + 1.28)/3 = 1.24$.

Likewise we have:

- Average seasonal index for second period in season $(1.23 + 1.24 + 1.28)/3 = 1.25$.
- Average seasonal index for third period in season $(0.51 + 0.77 + 0.86)/3 = 0.71$.
- Average seasonal index for fourth period in season $(0.67 + 0.83 + 0.88)/3 = 0.79$.

Now we have both the trend values and seasonal indices, so we can start forecasting for the future. For this we find the deseasonalised value in the future and then multiply this by the appropriate seasonal index.

## we Worked example 8.9

Forecast values for periods 13 – 17 for the time series in Worked Example 8.7.

## Solution

In Worked Example 8.7 we found the trend by linear regression:

trend value $= 223.41 + 18.05 \times$ period

Now we can substitute 13 to 17 for the period and find the deseasonalised trend values. Then we multiply these by the appropriate seasonal index – which we found in Worked Example 8.8 – to get the forecasts.

**Period 13**

Deseasonalised trend value $= 223.41 + 18.05 \times 13 = 458.06$

Seasonal index $= 1.24$   (first period in season)

Forecast $= 458.06 \times 1.24 = 568$

**Period 14**

Deseasonalised trend value $= 223.41 + 18.05 \times 14 = 476.11$

Seasonal index $= 1.25$   (second period in season)

Forecast $= 476.11 \times 1.25 = 595.14$

**Period 15**

Deseasonalised trend value $= 223.41 + 18.05 \times 15 = 494.16$

Seasonal index $= 0.71$   (third period in season)

Forecast $= 494.16 \times 0.71 = 350.85$

### Period 16

Deseasonalised trend value $= 223.41 + 18.05 \times 16 = 512.21$

Seasonal index $= 0.79$ (fourth period in season)

Forecast $= 512.21 \times 0.79 = 404.65$

### Period 17

Deseasonalised trend value $= 223.41 + 18.05 \times 17 = 530.26$

Seasonal index $= 1.24$ (first period in season)

Forecast $= 530.26 \times 1.24 = 657.52$

## we Worked example 8.10

Forecast values for the next four periods of the following time series.

| $t$ | 1 | 2 | 3 | 4 | 5 | 6 | 7 | 8 |
|-----|------|------|-----|-----|-----|------|-----|-----|
| $y$ | 986 | 1245 | 902 | 704 | 812 | 1048 | 706 | 514 |

## Solution

Looking at the data, we can see that there is a linear trend with a season of four periods. We could confirm this by drawing a graph, but the pattern is clear enough from the figures. We can use linear regression to deseasonalise the data, and then calculate four seasonal indices. Figure 8.14 shows these results in a spreadsheet.

This shows:

the linear regression equation: $y = 1156.75 - 64.92 \times \text{period}$

average seasonal indices: 0.94, 1.29, 0.97 and 0.8

The forecast for period 9 is:

$(1156.75 - 64.92 \times 9) \times 0.94 = 538.12$

| | A | B | C | D | E | F |
|---|---|---|---|---|---|---|
| 1 | **Forecasting** | | | | | |
| 2 | | | | | | |
| 3 | **Period** | **Observation** | **Regression** | **Seasonal** | **Period** | |
| 4 | | | **value** | **index** | **in season** | |
| 5 | 1 | 986 | 1091.83 | 0.90 | 1 | |
| 6 | 2 | 1245 | 1026.92 | 1.21 | 2 | |
| 7 | 3 | 902 | 962.00 | 0.94 | 3 | |
| 8 | 4 | 704 | 897.08 | 0.78 | 4 | |
| 9 | 5 | 812 | 832.08 | 0.98 | 1 | |
| 10 | 6 | 1048 | 767.25 | 1.37 | 2 | |
| 11 | 7 | 706 | 702.33 | 1.01 | 3 | |
| 12 | 8 | 514 | 637.42 | 0.81 | 4 | |
| 13 | | | | | | |
| 14 | **Summary results** | | | | | |
| 15 | | | | | | |
| 16 | *Regression statistics* | | | *Seasonal indices* | | |
| 17 | Correlation, r | 0.691 | | 1 | 0.94 | |
| 18 | Determination r^2 | 0.477 | | 2 | 1.29 | |
| 19 | Adjusted r^2 | 0.390 | | 3 | 0.97 | |
| 20 | Observations | 8 | | 4 | 0.8 | |
| 21 | | | | | | |
| 22 | *Coefficients* | | | | | |
| 23 | Intercept | 1156.75 | | | | |
| 24 | X variable 1 | −64.92 | | | | |
| 25 | | | | | | |
| 26 | | | | | | |
| 27 | **Forecasts** | | | | | |
| 28 | | | | | | |
| 29 | **Period** | **Trend** | **Seasonal** | **Value** | | |
| 30 | | **value** | **index** | | | |
| 31 | 9 | 572.47 | 0.94 | 538.12 | | |
| 32 | 10 | 507.55 | 1.29 | 654.74 | | |
| 33 | 11 | 442.63 | 0.97 | 429.35 | | |
| 34 | 12 | 377.71 | 0.8 | 302.17 | | |

**Figure 8.14**   Values for forecasts in Worked Example 8.10.

Similarly, the forecasts for periods 10, 11 and 12 are:

Period 10 :   $(1156.75 - 64.92 \times 10) \times 1.29 = 654.74$

Period 11:   $(1156.75 - 64.92 \times 11) \times 0.97 = 429.35$

Period 12:   $(1156.75 - 64.92 \times 12) \times 0.8 = 302.17$

## In summary

**The easiest way of forecasting complex time series is to consider separate components and combine them into a final forecast. A typical forecast comes from multiplying a trend value – found using linear regression – by a seasonal index.**

## Chapter review

This chapter discussed various aspects of forecasting. It started by considering the need to forecast and then described a number of forecasting methods. In particular it:

- outlined the importance of forecasting in all organisations;
- described three basic approaches to forecasting as causal (which we described by linear regression in Chapter 7), judgemental and projective;
- discussed the use of judgemental or qualitative methods when there is no relevant quantitative data;
- showed how time series can be described by an underlying pattern with superimposed random noise;
- forecast using actual averages, moving averages and exponential smoothing;
- produced forecasts for time series with seasonality and trend.

## Problems

8.1   Use linear regression to forecast values for periods 11 to 13 with the following time series.

| Period | 1 | 2 | 3 | 4 | 5 | 6 | 7 | 8 | 9 | 10 |
|---|---|---|---|---|---|---|---|---|---|---|
| Observation | 121 | 133 | 142 | 150 | 159 | 167 | 185 | 187 | 192 | 208 |

8.2   Use actual averages to forecast values for the data in Problem 8.1. Which method gives the better results?

8.3   Use a four-period moving average to forecast values for the data in Problem 8.1.

8.4   Find the two-, three- and four-period moving averages for the following time series, and see which gives the best results.

| $t$ | 1 | 2 | 3 | 4 | 5 | 6 | 7 | 8 |
|---|---|---|---|---|---|---|---|---|
| $y$ | 280 | 240 | 360 | 340 | 300 | 220 | 200 | 360 |

8.5 Use exponential smoothing with $\alpha = 0.1$ and 0.2 to forecast values for the data in Problem 8.4. Which gives the better forecast?

8.6 Use exponential smoothing with smoothing constant $\alpha$ equal to 0.1, 0.2, 0.3 and 0.4 to get forecasts for one period ahead for the following time series. Use an initial value of $F_1 = 208$ and say which value of $\alpha$ is best.

| $t$ | 1 | 2 | 3 | 4 | 5 | 6 | 7 | 8 |
|-----|-----|-----|-----|-----|-----|-----|-----|-----|
| $y$ | 212 | 216 | 424 | 486 | 212 | 208 | 208 | 204 |

8.7 Forecast values for the next three periods of the following time series.

| $t$ | 1 | 2 | 3 | 4 | 5 | 6 |
|-----|-----|-----|-----|-----|-----|-----|
| $y$ | 100 | 160 | 95 | 140 | 115 | 170 |

8.8 Forecast values for the next eight periods of the following time series.

| $t$ | 1 | 2 | 3 | 4 | 5 | 6 | 7 | 8 | 9 | 10 |
|-----|-----|-----|-----|-----|-----|-----|-----|-----|-----|-----|
| $y$ | 101 | 125 | 121 | 110 | 145 | 165 | 160 | 154 | 186 | 210 |

# CS Case study – Workload planning

Mary James was in charge of the purchasing department of Castle Town Construction. One morning she walked into the office and said: 'The problem with this office is lack of planning. I have been reading a few articles and it seems to me that forecasting is the key to an efficient business. We never do any forecasting, but simply rely on experience to guess our future workload. I think we should start using exponential smoothing.'

Unfortunately, the office was busy and nobody had time to look at forecasting. A month later nothing had actually happened. To make some progress with the forecasting, Mary got a management trainee to work on some figures. The trainee examined their work, and divided it into seven categories, including searching for business, preparing estimates, submitting tenders, finding suppliers, and so on. For each of these categories he found the number of jobs that the office had done in each quarter of the past three years. His data are shown overleaf.

| Quarter | Work 1 | Work 2 | Work 3 | Work 4 | Work 5 | Work 6 | Work 7 |
|---------|--------|--------|--------|--------|--------|--------|--------|
| 1 | 129 | 74 | 1000 | 755 | 1210 | 204 | 24 |
| 2,1 | 138 | 68 | 1230 | 455 | 1520 | 110 | 53 |
| 3,1 | 110 | 99 | 890 | 810 | 1390 | 105 | 42 |
| 4,1 | 118 | 119 | 700 | 475 | 1170 | 185 | 21 |
| 1,2 | 121 | 75 | 790 | 785 | 1640 | 154 | 67 |
| 2,2 | 137 | 93 | 1040 | 460 | 1900 | 127 | 83 |
| 3,2 | 121 | 123 | 710 | 805 | 1860 | 187 | 80 |
| 4,2 | 131 | 182 | 490 | 475 | 1620 | 133 | 59 |
| 1,3 | 115 | 103 | 610 | 775 | 2010 | 166 | 105 |
| 2,3 | 126 | 147 | 840 | 500 | 2340 | 140 | 128 |
| 3,3 | 131 | 141 | 520 | 810 | 2210 | 179 | 126 |
| 4,3 | 131 | 112 | 290 | 450 | 1990 | 197 | 101 |

Now the trainee wanted to forecast likely workload for the next two years. For this he found the number of 'standard work units' that were needed for a typical job in each type of work:

Work 1     2 units

Work 2     1.5 units

Work 3     1 unit

Work 4     0.7 units

Work 5     0.4 units

Work 6     3 units

Work 7     2.5 units

What do you think the trainee should do now? How useful do you think the results will be?

# 9 Some other types of problem

## Contents

## Chapter outline

In this chapter we are going to look at some other types of quantitative model. In particular, we will outline problems of inventory control, linear programming and network analysis. After reading this chapter you should be able to:

- understand both the need for stocks and the associated costs
- calculate economic order quantities and reorder levels
- understand the idea of constrained optimisation
- formulate linear programmes
- interpret the solutions to linear programmes
- appreciate the need to plan complex projects
- draw networks to represent projects
- find the timing of projects

# 9.1 Background to stock control

Every organisation holds stocks. These are the stores of goods that are kept until they are needed. Keeping stock is expensive – with costs of warehousing, tied-up capital, deterioration, loss and so on. So we can start by asking, 'why do organisations hold stock?' The main answer is that stocks give a buffer between variable supply and demand. Imagine a supermarket that has a lorry making one large delivery every day, while its customers have small demands that occur almost continuously: there is a mismatch between supply and demand. The supermarket overcomes this mismatch by holding stock.

> The main purpose of **stock** is to give a buffer between supply and demand.

There are other reasons for holding stock, including:

- to act as a buffer between different production operations (it 'decouples' operations);
- to allow for demands that are larger than expected, or at unexpected times;
- to allow for deliveries that are delayed or too small;
- to take advantage of price discounts on large orders;
- to buy items when the price is low and expected to rise;
- to buy items that are going out of production or are difficult to find;
- to reduce transport costs by making up full loads;
- to give cover for emergencies.

Whatever the reason for holding stock, there are associated costs and these can be surprisingly high. Organisations look for policies that minimise the costs of holding stock; in particular, they look for answers to three basic questions.

- What items should they hold in stock?
- When should they place orders to replenish stock?
- How much should they order at a time?

Here we will concentrate on the last two questions. To answer these, we need a clearer picture of the costs, and will define four distinct types:

- **Unit Cost** $(U_c)$  This is the price of an item charged by the supplier, or the cost to the company of acquiring one unit of an item. It may be fairly easy to find this by looking at quotations or recent invoices, but sometimes it is more difficult – especially when there are several suppliers, a choice of products, and different purchase conditions. If a company makes the item itself, it can be difficult to set a reasonable production or transfer cost.

- **Reorder Cost** $(R_c)$    This is the cost of placing a repeat order for an item and might include allowances for drawing up an order, computer processing, correspondence and telephone costs, receiving, follow-up action and so on. The reorder cost may also include some quality control, transport charges, sorting and movement of goods received.

- **Holding Cost** $(H_c)$    This is the cost of holding one unit of an item in stock for a period of time. The obvious cost is tied-up money which is either borrowed or could be put to other uses. Other holding costs cover storage space, deterioration, obsolescence, handling, administration and insurance.

- **Shortage Cost** $(S_c)$    If an item is needed but cannot be supplied from stock, there is a shortage cost. In the simplest case a retailer may lose direct profit from a sale, but the effects of shortages are usually much wider than this. There might also be a cost of lost goodwill, potential future sales, and loss of reputation. Shortages of raw materials might disrupt production, and there might also be costs for positive action such as sending out emergency orders, paying for special deliveries, storing partly finished goods or using alternative, more expensive suppliers.

    Shortage costs are usually difficult to find, but they can be very high, particularly if production is stopped by a shortage of raw materials. This means that we can look at the purpose of stock again and say, 'the cost of shortages can be very high and to avoid these organisations are willing to pay the relatively lower costs of carrying stock'.

### In summary

**The main purpose of stock is to act as a buffer between supply and demand. Holding stock is expensive. Organisations look for policies that minimise the total costs of their stock.**

# 9.2 Economic order quantity

## 9.2.1 Developing the model

Imagine a single item that is held in stock to meet a demand that is constant at $D$ per unit time. When stock is ordered it arrives immediately, and we know exactly the unit cost $(U_c)$, reorder cost $(R_c)$ and holding cost $(H_c)$. The shortage cost $(S_c)$ is so high that all demands must be met and no shortages are allowed. Using this information we can find an **economic order quantity**, which is the best order size. We will always order this amount, so the stock level rises with deliveries and falls as units are removed to meet demand, giving the sawtooth pattern shown in Figure 9.1.

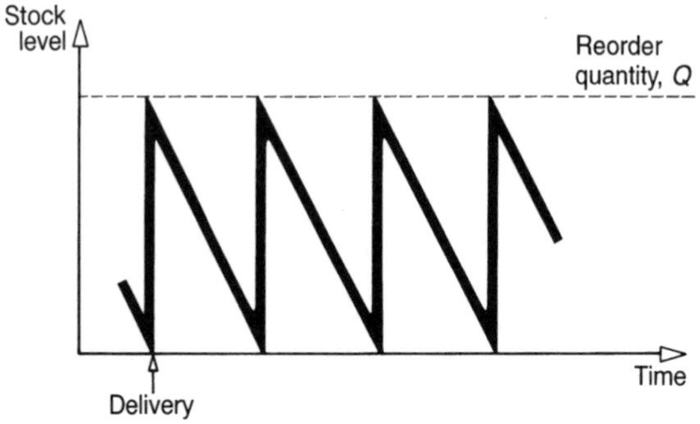

**Figure 9.1**   Sawtooth stock level over time.

At regular periods an amount $Q$ arrives, and then the stock declines at a constant rate, $D$, until there is no stock left. Then another order arrives. The resulting stock cycle has a length $T$. A standard result shows that the best order size is the economic order quantity, $Q_o$, which is:

$$\text{economic order quantity} = Q_o = \sqrt{\frac{2R_cD}{H_c}}$$

This result gives the minimum cost of holding stock. This cost has two parts: a fixed part, $U_cD$, which does not vary with order quantity, and a variable part, $VC_o$, which does. Optimal values for these are:

$$\text{total cost} = TC_o = U_cD + VC_o$$

$$\text{variable cost} = VC_o = \sqrt{2R_cH_cD}$$

## we Worked example 9.1

The demand for an item is constant at 20 units a month. The unit cost is £50, the cost of processing an order is £60, and the holding cost is £18 per unit per year. What are the economic order quantity, cycle length and costs?

## Solution

Listing the values that we know in consistent units:

$D = 20 \times 12 = 240$ units per year

$U_c = £50$ per unit

$R_c = £60$ per order

$H_c = £18$ per unit per year

We can substitute these values to give:

$$Q_o = \sqrt{\frac{2R_cD}{H_c}} = \sqrt{\frac{2 \times 60 \times 240}{18}} = 40 \text{ units}$$

$$VC_o = \sqrt{2R_cH_cD} = \sqrt{2 \times 60 \times 18 \times 240} = £720 \text{ per year}$$

$$TC_o = U_c\,D + VC_o = (50 \times 240) + 720 = £12\,720 \text{ per year}$$

The amount of stock delivered during a cycle is $Q$, and the amount leaving is $DT$, where $T$ is the length cycle. These must be equal, so $Q = DT$. The optimal values are

$$Q_o = DT_o$$

so

$$40 = 240 \times T_o$$

giving

$$T_o = \tfrac{1}{6} \text{ years} = 2 \text{ months}$$

Hence, optimal policy (with total costs of £12 720 a year) is to order 40 units every 2 months.

## In summary

**After making some assumptions, we can calculate an economic order quantity. This is the order size that minimises the total cost of stocking an item.**

### 9.2.2  Reorder levels with fixed lead times

The economic order quantity answers the question of how much to order, but we still need to know when to place an order. To answer this we have to look at the leadtime, $L$, between placing an order and having it arrive in stock. For simplicity, we will assume that $L$ is fixed.

We know that the stock level follows the sawtooth pattern shown in Figure 9.1. To make sure that a delivery arrives just as stock is running out, we must place an order a time $L$ earlier. The easiest way of finding this point is to look at the current stock and place an order when there is just enough left to last the leadtime. With constant demand of $D$, this means that we place an order when the stock level falls to $LD$. This point is called the **reorder level** (shown in Figure 9.2).

> reorder level = leadtime demand
>
> $ROL = LD$

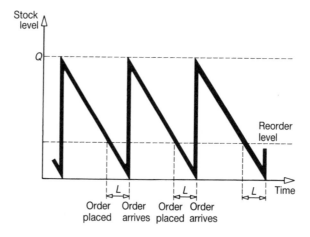

Order Order    Order Order
placed arrives placed arrives

**Figure 9.2**    Stock level with a fixed leadtime *L*.

One way of timing orders in practice is called the 'two-bin system'. This system sees stock kept in two bins: the first bin holds the reorder level while the second bin holds all remaining stock. Demand is met from the second bin until this is empty. At this point the stock level has fallen to the reorder level and it is time to place another order.

## we Worked example 9.2

Demand for an item is constant at 20 units a week, the reorder cost is £125 an order, and the holding cost is £2 per unit per week. If suppliers guarantee delivery within 2 weeks, what is the best ordering policy?

## Solution

Listing the variables in consistent units:

$D$ = 20 units per week

$R_c$ = £125 per order

$H_c$ = £2 per unit per week

$L$ = 2 weeks

We can substitute these values to get:

Economic order quantity: $Q_o = \sqrt{\dfrac{2R_cD}{H_c}} = \sqrt{\dfrac{2 \times 125 \times 20}{2}} = 50$ units

Reorder level: $\qquad\qquad ROL = LD = 2 \times 20 = 40$ units

The best policy is to place an order for 50 units whenever stock declines to 40 units. We can find the cycle length, $T_o$, from:

$$Q_o = DT_o$$

so

$$T_o = \frac{50}{20} = 2.5 \text{ weeks}$$

The variable cost is:

$$VC_o = \sqrt{2R_cH_cD} = \sqrt{2 \times 125 \times 2 \times 20} = £100 \text{ a week}$$

## In summary

**The reorder level indicates the time to place an order. For constant leadtime and demand, the reorder level equals leadtime demand.**

# 9.3 Introduction to linear programming

Managers often want to achieve some objective, but find that there are constraints on the amount of resources available. An operations manager wants to maximise the number of units made using limited production facilities; a marketing manager wants to maximise the impact of an advertising campaign without spending more than a specified budget; a finance manager wants to maximise the return on investment of a limited amount of funds; a construction manager wants to minimise the cost of a project without exceeding the time available, and so on. Such problems have:

- an objective of optimising (i.e. maximising or minimising) some function;

- a set of constraints that limit the possible solutions.

For this reason they are called problems of **constrained optimisation**.

**Linear programming** (LP) is a widely used method of solving problems of constrained optimisation – and it has nothing to do with computer programming. There are three steps in solving a linear programme:

- **formulation**, to get the problem into the right form;

- **solution**, to find an optimal solution to the problem;

- **analysis**, interpreting the solution.

In practice, formulation is the most difficult step. When a problem is in the right form, getting a solution is quite straightforward – because we always use a computer.

## In summary

**Many problems in business can be described as constrained optimisation. Linear programming is a widely used method of tackling such problems.**

# 9.4 Getting LP problems into the right form

We will show how to formulate problems using an example. Suppose a small factory makes two types of liquid fertiliser: Growbig and Thrive. These are made by similar processes, using the same equipment for blending raw materials, distilling the mix and finishing (bottling, testing, weighing, etc.). Because the factory has a limited amount of equipment, there are constraints on the time available for each process. In particular, no more than 40 hours of blending is available in a week, 40 hours of distilling and 25 hours of finishing.

    The fertilisers are made in batches of 1000 litres and each batch needs the following number of hours on each process:

|  | Growbig | Thrive |
|---|---|---|
| Blending | 1 | 2 |
| Distilling | 2 | 1 |
| Finishing | 1 | 1 |

    If the factory makes a net profit of £30 on each batch of Growbig and £20 on each batch of Thrive, how many batches of each should it make a week?

    This is a problem of constrained optimisation, where we want to maximise the profits subject to constraints on production (Figure 9.3). We could use common sense to say that the profit on Growbig is higher than the profit on Thrive, so we should make as much Growbig as possible. Then we can make 20 batches of Growbig before all the distilling time is used, and this gives a profit of £600. However this is not the best answer.

    We want to find the optimal number of batches of Growbig and Thrive, so our first step is to define these as the **decision variables**:

Let $G$ be the number of batches of Growbig made in a week

Let $T$ be the number of batches of Thrive made in a week

Now look at the time available for blending. Each batch of Growbig uses 1 hour of blending, so $G$ batches use $G$ hours; each batch of Thrive uses 2 hours of blending, so $T$ batches use $2T$ hours. Adding these together gives the total amount of blending time used as $G + 2T$. The maximum amount of blending time available is 40 hours, so the time used must be less than – or at worst equal to – this. So we have the first constraint:

$$G + 2T \leq 40 \qquad \text{blending constraint}$$

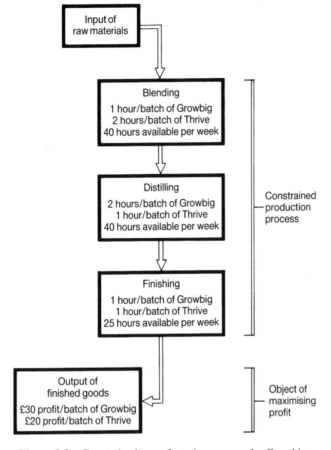

**Figure 9.3**  Constrained manufacturing process for Growbig and Thrive.

(Remember that ≤ means less than or equal to.) Now look at the distilling constraint. Each batch of Growbig uses 2 hours of distilling, so $G$ batches use $2G$ hours; each batch of Thrive uses 1 hour of distilling, so $T$ batches use $T$ hours. Adding these together gives the total amount of distilling used, and this must be less than, or equal to, the amount of distilling available (40 hours). This gives the second constraint:

$$2G + T \le 40 \qquad \text{distilling constraint}$$

Now the finishing constraint has the total time used for finishing ($G$ for batches of Growbig plus $T$ for batches of Thrive) less than or equal to the time available (25 hours) to give:

$$G + T \le 25 \qquad \text{finishing constraint}$$

These are the three constraints for the process, but there is another implicit condition. The company cannot make a negative number of batches, so both $G$ and $T$ must be positive:

$$G \geq 0 \quad \text{and} \quad T \geq 0 \qquad \text{non-negativity constraints}$$

This **non-negativity constraint** is a standard feature of linear programmes.

Now we have all the constraints, and can turn to the objective. The objective is to maximise profit. We make £30 on each batch of Growbig, so $G$ batches make a profit of $30G$; we make £20 on each batch of Thrive, so $T$ batches make a profit of $20T$. Adding these gives the total profit to be maximised. This is called the **objective function**:

$$\text{Maximise:} \quad 30G + 20T \qquad \text{objective function}$$

Now the linear programming formulation is complete with descriptions of:

- decision variables
- an objective function
- a set of constraints
- non-negativity constraints

In this example we have:

- $G$ and $T$                             decision variables
- Maximise    $30G + 20T$           objective function
- Subject to:   $G + 2T \leq 40$          constraints
  
                  $2G + T \leq 40$
  
                    $G + T \leq 25$
- $G \geq 0$ and $T \geq 0$           non negativity constraints

## we Worked example 9.3

A political campaign wants to hire photocopying machines to make leaflets for a local election. There are two suitable machines:

- Acto costs £120 a month to rent, occupies 2.5 square metres of floorspace and can make 15 000 copies a day.

- Zenmat costs £150 a month to rent, occupies 1.8 square metres of floorspace and can make 18 500 copies a day.

The campaign has allowed up to £1200 a month for copying machines which will be put in a room of 19.2 square metres. How would you formulate this problem?

## Solution

The problem variables are the things that we can vary, which are the number of Acto and Zenmat machines rented, so for our decision variables:

Let $A$ be the number of Acto machines rented

Let $Z$ be the number of Zenmat machines rented

The objective is to make as many copies as possible:

Maximise:   $15\,000A + 18\,500Z$      objective function

There are constraints on floorspace and costs, so:

Subject to:   $120A + 150Z \le 1200$      cost constraint

$2.5A + 1.8Z \le 19.2$      space constraint

$A \ge 0$ and $Z \ge 0$      non-negativity constraint

## we Worked example 9.4

An investment trust has £1 million to invest. After consulting its financial advisers it decides that there are six possible investments (Table 9.1).

**Table 9.1**

| Investment | % risk | % dividend | % growth | Rating |
|---|---|---|---|---|
| 1 | 18 | 4 | 22 | 4 |
| 2 | 6 | 5 | 7 | 10 |
| 3 | 10 | 9 | 12 | 2 |
| 4 | 4 | 7 | 8 | 10 |
| 5 | 12 | 6 | 15 | 4 |
| 6 | 8 | 8 | 8 | 6 |

The trust wants to invest the £1 million with minimum risk, but with a dividend of at least £70 000 a year, average growth of at least 12% and average rating of at least 7. Formulate this problem as a linear programme.

## Solution

The decision variables are the amount of money put into each investment.

Let $X_1$ be the amount of money put into investment 1

Let $X_2$ be the amount of money put into investment 2, etc.

So $X_i$ is the amount of money put into investment $i$.

The objective is to minimise risk.

Minimise:   $0.18X_1 + 0.06X_2 + 0.10X_3 + 0.04X_4 + 0.12X_5 + 0.08X_6$

There are constraints on the following:

Money – the total invested must equal £1 million:

$$X_1 + X_2 + X_3 + X_4 + X_5 + X_6 = 1\,000\,000$$

Dividend – which must be at least 7% of £1 million:

$$0.04X_1 + 0.05X_2 + 0.09X_3 + 0.07X_4 + 0.06X_5 + 0.08X_6 \geq 70\,000$$

Average growth – which must be at least 12% of £1 million:

$$0.22X_1 + 0.07X_2 + 0.12X_3 + 0.08X_4 + 0.15X_5 + 0.08X_6 \geq 120\,000$$

Rating – the average, weighted by the amount invested, must be at least 7:

$$4X_1 + 10X_2 + 2X_3 + 10X_4 + 4X_5 + 6X_6 \geq 7\,000\,000$$

The non-negativity constraints $X_1, X_2, X_3, X_4, X_5$ and $X_6 \geq 0$ complete the formulation.

## In summary

**The first stage in solving a linear programming is to describe a problem in a standard form. This formulation contains decision variables, an objective function and constraints.**

## 9.5 Solving linear programmes

Finding the solution to a linear programme needs a lot of arithmetic, and you will always do this with a computer. LP packages vary in detail, but they all give essentially the same information. Figure 9.4 shows a typical printout for the problem of Growbig and Thrive described above.

The first part of this printout shows the data to confirm that they were entered properly. The next part shows the main results. The optimal solution is:

$G = 15$ batches

$T = 10$ batches

Profit = £650

The **original coefficient** is the profit made on each unit of $G$ and $T$. If the optimal solution gives no units of $G$ or $T$ made, the **coefficient sensitivity** shows how much the profit on a unit must rise before it joins the optimal solution.

```
                    -=*=-  INFORMATION ENTERED  -=*=-

NUMBER OF VARIABLES             :    2
NUMBER OF <= CONSTRAINTS        :    3
NUMBER OF  = CONSTRAINTS        :    0
NUMBER OF >= CONSTRAINTS        :    0
MAXIMISE Profit =   30   G   +    20   T
SUBJECT TO:
Blending      1   G   +   2   T        <=   40
Distilling    2   G   +   1   T        <=   40
Finishing     1   G   +   1   T        <=   25
```

                        -=*=-  RESULTS  -=*=-

| VARIABLE | VARIABLE VALUE | ORIGINAL COEFFICIENT | COEFFICIENT SENSITIVITY |
|---|---|---|---|
| G | 15 | 30 | 0 |
| T | 10 | 20 | 0 |

| CONSTRAINT | ORIGINAL RIGHT-HAND VALUE | SLACK OR SURPLUS | SHADOW PRICE |
|---|---|---|---|
| Blending | 40 | 5 | 0 |
| Distilling | 40 | 0 | 10 |
| Finishing | 25 | 0 | 10 |

OBJECTIVE FUNCTION VALUE:         650

                    -=*=-  SENSITIVITY ANALYSIS  -=*=-

OBJECTIVE FUNCTION COEFFICIENTS

| VARIABLE | LOWER LIMIT | ORIGINAL COEFFICIENT | UPPER LIMIT |
|---|---|---|---|
| G | 20 | 30 | 40 |
| T | 15 | 20 | 30 |

RIGHT-HAND SIDE VALUES

| CONSTRAINT | LOWER LIMIT | ORIGINAL VALUE | UPPER LIMIT |
|---|---|---|---|
| Blending | 35 | 40 | NO LIMIT |
| Distilling | 35 | 40 | 50 |
| Finishing | 20 | 25 | 26.667 |

                --------  END OF ANALYSIS  --------

**Figure 9.4**  Printout from LP package for Growbig/Thrive problem.

The next part shows the amount of each constraint available. The **slack or surplus** shows how much spare capacity there is with the optimal solution – so there is only 5 hours of spare capacity in blending. The **shadow price** shows how much each extra unit of resource raises the profit. Blending already has spare capacity, so an extra unit has no value, but each extra unit of either distilling or finishing raises the profit by £10.

The next part of the printout shows a sensitivity analysis for the objective function. The original profit on each batch of Growbig is £30 but this can vary between £20 and £40 without changing the optimal values of $G$ and $T$. Similarly, the profit on each batch of Thrive can vary between £15 and £30 without changing the optimal values.

The final table shows the range in which the shadow prices are valid. The distilling constraint has a shadow price of £10 provided there is between 35 and 50 hours available. Similarly, the shadow price of blending is zero provided there is more than 35 hours available.

## we Worked example 9.5

West Coast Wood Products makes four types of pressed panel from pine and spruce. Each sheet of panel must be cut and pressed. The following table shows the hours needed to make a batch of each type of panel, and the hours available each week.

**Table 9.2**

| Panel type | Hours of cutting | Hours of pressing |
|---|---|---|
| Classic | 1 | 1 |
| Western | 1 | 4 |
| Nouveau | 2 | 3 |
| East Coast | 2 | 2 |
| **Available** | **80** | **100** |

There is a limited amount of suitable wood. The amounts needed for a batch of each type of panel and maximum weekly availability are given in Table 9.3.

**Table 9.3**

| | Classic | Western | Nouveau | East Coast | Available |
|---|---|---|---|---|---|
| Pine | 50 | 40 | 30 | 40 | **2500** |
| Spruce | 20 | 30 | 50 | 20 | **2000** |

The profit on each batch of panels is £40 for Classic, £110 for Western, £75 for Nouveau and £35 for East Coast.

(a) Formulate this as a linear programme.

(b) A computer program gives the results for this problem shown in Figure 9.5. What do these show?

```
            − = * = −    INFORMATION ENTERED    − = * = −

NUMBER OF VARIABLES          :  4
NUMBER OF <= CONSTRAINTS     :  4
NUMBER OF  = CONSTRAINTS     :  0
NUMBER OF >= CONSTRAINTS     :  0

MAXIMISE Profit =   40   CLS  +   110   WST  +   75   NOU  +   35   EST

SUBJECT TO:

   50   CLS  +   40   WST  +   30   NOU  +   40   EST  <=   2500
   20   CLS  +   30   WST  +   50   NOU  +   20   EST  <=   2000
    1   CLS  +    1   WST  +    2   NOU  +    2   EST  <=     80
    1   CLS  +    4   WST  +    3   NOU  +    2   EST  <=    100


                 − = * = −    RESULTS    − = * = −

                       VARIABLE        ORIGINAL        COEFFICIENT
       VARIABLE         VALUE         COEFFICIENT      SENSITIVITY

         CLS            37.5              40                0
         WST           15.625            110               0
         NOU             0               75               7.5
         EST             0               35              26.25


    CONSTRAINT        ORIGINAL        SLACK OR          SHADOW
     NUMBER        RIGHT-HAND VALUE    SURPLUS           PRICE

        1              2500              0               0.312
        2              2000            781.25             0
        3               80            26.875             0
        4               100              0              24.375


OBJECTIVE FUNCTION VALUE:          3218.75

                 − −    SENSITIVITY ANALYSIS    − −

                                    OBJECTIVE FUNCTION COEFFICIENTS

                      LOWER          ORIGINAL           UPPER
       VARIABLE       LIMIT         COEFFICIENT         LIMIT
         CLS          27.5              40              137.5
         WST          100              110              160
         NOU        NO LIMIT           75              82.5
         EST        NO LIMIT           35              61.25


                              RIGHT-HAND SIDE VALUES

    CONSTRAINT        LOWER          ORIGINAL           UPPER
     NUMBER           LIMIT           VALUE             LIMIT
        1             1000            2500            3933.333
        2            1218.75          2000            NO LIMIT
        3            53.125            80             NO LIMIT
        4              50              100              250


         − − − − − − −    END OF ANALYSIS    − − − − − − − −
```

**Figure 9.5**   Printout for Worked Example 9.5.

## Solution

(a)  The decision variables are the number of batches of each type of panel made a week (CLS, WST, NOU and EST). Then the formulation is:

Maximise:   $40 \times \text{CLS} + 110 \times \text{WST} + 75 \times \text{NOU} + 35 \times \text{EST}$

Subject to:   $50 \times \text{CLS} + 40 \times \text{WST} + 30 \times \text{NOU} + 40 \times \text{EST} \leq 2500$   pine

$20 \times \text{CLS} + 30 \times \text{WST} + 50 \times \text{NOU} + 20 \times \text{EST} \leq 2000$   spruce

$1 \times \text{CLS} + 1 \times \text{WST} + 2 \times \text{NOU} + 2 \times \text{EST} \leq 80$   cutting

$1 \times \text{CLS} + 4 \times \text{WST} + 3 \times \text{NOU} + 2 \times \text{EST} \leq 100$   pressing

with CLS, WST, NOU and EST $\geq 0$

(b)  The results in Figure 9.5 show:

- the optimal solution is to make 37.5 batches of Classic a week, 15.625 batches of Western and none of the others;

- the profit is £3218.75;

- the profit on batches of Nouveau and East Coast would have to rise by £7.50 and £26.25 respectively before the optimal solution includes these panels;

- the limiting constraints are pine and pressing (numbers 1 and 4), with spare capacity in spruce (781.25) and cutting (26.875 hours). This will remain true while the constraint on spruce remains over 1218.75 and the amount of cutting remains above 53.13 hours;

- extra units of pine would raise the profit by £0.312 each (valid for amounts between 1000 and 3933.333) and of pressing is £24.375 (valid for between 50 and 250 hours);

- the profit for each batch of Classic can vary between £27.50 and £137.50 without changing the optimal values of the variables – with similar ranges of £100.00 to £160.00 for Western, below £82.50 for Nouveau and below £61.25 for East Coast.

---

Linear programmes often have thousands of variables and constraints, so it is not surprising that there are sometimes mistakes in the formulation. Two particular concerns are:

- **unbound solution**, which means that the constraints do not limit the solution, and some decision variables become infinite;

- **infeasible solution**, which means that the constraints are too tight and there is no feasible solution.

If you get either of these, you have to check the input data and find the mistake.

## In summary

**You will always use a computer to solve a linear programme. The printouts differ in detail, but they give a range of standard information.**

# 9.6 Project network analysis

A **project** is a coherent piece of work with a distinct start and finish.

We all do a number of small projects every day, such as preparing a meal, writing a report, building a fence or organising a social function. We have to plan these projects, and in particular find:

- the activities that make up the project;
- the order in which these activities have to be done;
- the timing of each activity;
- the resources needed at each stage.

Small projects need almost no formal planning but, in business, a project can be much bigger and more complicated – like the installation of a new computer system, building a nuclear power station, organising the Olympic Games, and building a rail tunnel under the English Channel. These large projects will only run smoothly if they are carefully planned. The most widely used method of organising large projects is **project network analysis**, which is described in the rest of this chapter.

## In summary

**A project is a coherent piece of work that has a clear start and finish. Projects can be very big and need careful planning. Project network analysis is the most widely used method of doing this planning.**

# 9.7 Drawing networks for projects

## 9.7.1 Drawing networks

A project network consists of a series of nodes connected by arrows. Each arrow represents an activity in the project, and each node represents the point at which activities begin and end. The nodes are called **events** and a network consists of alternating activities and events.

Figure 9.6 shows part of a project network which has two activities, A followed by B. Event 1 is the start of activity A, event 2 is the finish of activity A and the start of activity B, and event 3 is the finish of activity B. The arrows show only relationships between activities and there is no significance in their direction or length.

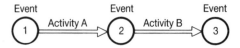

**Figure 9.6** Part of project network.

We can draw networks for much larger and more complicated projects.

# we Worked example 9.6

Jane is going to build a greenhouse from a kit. The instructions show that this is a project with three parts:

A, preparing the base (which will take 3 days);

B, building the frame (which will take 2 days);

C, fixing the glass (which will take 1 day).

Draw a network for the project.

## Solution

The project has three activities which must be done in a fixed order; building the frame must be done after preparing the base and before fixing the glass. We can describe this order by a **dependency table**. This lists each activity along with those activities that must immediately precede it (Table 9.4).

**Table 9.4**

| Activity | Duration (days) | Description | Immediate predecessor |
|----------|-----------------|-------------|-----------------------|
| A | 3 | Prepare base | – |
| B | 2 | Build frame | A |
| C | 1 | Fix glass | B |

Labelling the activities A, B and C is a convenient shorthand so that we can refer to activity B having activity A as immediate predecessor – which is normally expressed as 'B depends on A'. This table gives only immediate predecessors, so the fact that activity C (fixing the glass) depends on activity A as well as B need not be shown separately but can be inferred from other dependencies. Activity A has no immediate predecessor and can be started whenever convenient.

Figure 9.7 shows a network for this dependency table.

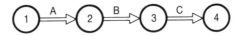

**Figure 9.7**    Network for Worked Example 9.6.

After drawing the network for a project we can look at the timing. For this we use a notional starting time of 0, and then calculate the start and finish times of each activity.

# we Worked example 9.7

Find the times for each activity in the project described in Worked Example 9.6. What happens if preparing the base takes more than three days, or building the frame takes less than two days?

## Solution

If we take a starting time of 0, then preparing the base can be finished by the end of day 3. Then building the frame can start, and, as it takes two days, can be finished by the end of day 5. Then fixing the glass can start; as this takes one day, it can be finished by the end of day 6.

If the base takes more than three days the project will be delayed. If the frame take less than two days the project will be finished early.

We now have a timetable for the project showing when each activity starts and finishes. We can use this timetable to schedule resources – such as hiring a concrete mixer for building the base of the greenhouse. We can list the steps in project planning as:

- Define the separate activities and their durations.
- Find the dependencies of activities.
- Draw a network.
- Analyse the timing of the project.
- Schedule resources.

We can extend these ideas to draw networks of almost any size. A useful approach is to start drawing the network on the left-hand side with those activities that do not depend on any other. Then we can add activities that depend only on these first activities, then we can add the activities that depend only on the latest activities, and so on. In this way we can expand the network systematically working from left to right until we have added all activities and the network is complete.

There are two important rules for drawing networks:

● Each project has a single start and finish node.

● Only one activity goes between any two nodes.

## we Worked example 9.8

A company is opening a new office and identifies the main activities and dependencies in Table 9.5. Draw a network of this project.

**Table 9.5**

| Activity | Description | Depends on |
|----------|-------------|------------|
| A | Find office location | – |
| B | Recruit new staff | – |
| C | Make office alterations | A |
| D | Order equipment needed | A |
| E | Install new equipment | D |
| F | Train staff | B |
| G | Start operations | C, E, F |

## Solution

Activities A and B have no predecessors and can be started as soon as convenient. As soon as activity A is finished, both C and D can start: E can start as soon as D is finished and F can start as soon as B is finished. G can start only when C, E and F have all finished. This gives the network shown in Figure 9.8.

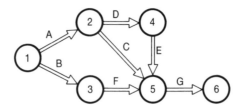

**Figure 9.8**   Network for Worked Example 9.8.

## In summary

**Project network analysis divides a project into separate activities. A dependency table shows the order of the activities, and these can be drawn in a network of alternating activities and events. After drawing the network we can do some calculations for the timing and resource allocation.**

### 9.7.2 Dummy activities

There are two circumstances which make drawing networks a bit more complicated. Table 9.6 shows the first of these.

**Table 9.6**

| Activity | Depends on |
|----------|-----------|
| A | – |
| B | A |
| C | A |
| D | B, C |

You might be tempted to draw this network like Figure 9.9(a), but this has two activities going between two events, which breaks the second rule above. The way around this is to use a **dummy activity**. This is not a part of the project, has zero duration and uses no resources, but is simply there to give a sensible network. In this case the dummy makes sure that only one activity goes between two events and is called a **uniqueness dummy**. Figure 9.9(b) shows the dummy activity as the broken line, X.

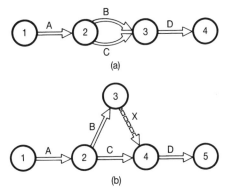

**Figure 9.9**   Network with uniqueness dummy: (a) incorrect network; (b) correct network using dummy activity X.

Table 9.7 shows part of a dependency table with a second situation that needs a dummy activity.

**Table 9.7**

| Activity | Depends on |
|----------|-----------|
| D | not given |
| E | not given |
| F | D, E |
| G | D |

You might be tempted to draw this part of the network as shown in Figure 9.10(a), but the dependency is clearly wrong. Activity F is shown as depending on D and E, which is correct – but G has the same dependency. The dependency table shows that G can start as soon as D is finished, but the network shows it waiting for E to finish as well. To get around this, we separate the dependencies using a dummy activity, as show in Figure 9.10(b). The dependence of F on D is shown through the dummy activity X. In effect, the dummy cannot start until D has finished and then F cannot start until the dummy and E are finished: as the dummy activity has zero duration this does not add any time to the project. This type of dummy is called a **logical dummy**.

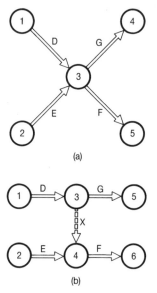

(a)

(b)

**Figure 9.10**   Network with a logical dummy: (a) incorrect network;
(b) correct network using dummy activity X.

## we Worked example 9.9

A project is described by the dependency table of Table 9.8. Draw a network of the project.

**Table 9.8**

| Activity | Depends on | Activity | Depends on |
|----------|-----------|----------|-----------|
| A | J | I | J |
| B | C, G | J | – |
| C | A | K | B |
| D | F, K, N | L | I |
| E | J | M | I |
| F | B, H, L | N | M |
| G | A, E, I | O | M |
| H | G | P | O |

## Solution

This seems a difficult network, but the steps are fairly straightforward. Activity J is the only one that does not depend on anything else, so this starts the network. Then we add activities A, E and I, which depend only on J. Then we can add activities which depend on A, E and I. Continuing this systematic addition of activities leads to the network shown in Figure 9.11. This includes four dummy activities, W, X, Y and Z.

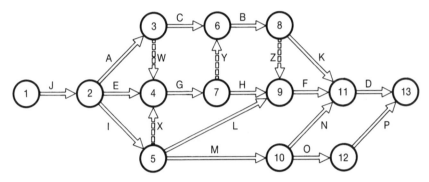

**Figure 9.11**   Network for Worked Example 9.9.

### In summary

**We may need two types of dummy activity to draw a network. Uniqueness dummies make sure that only one activity starts and finishes with the same events; logical dummies make sure that the logic of the network is accurate.**

# 9.8 Timing of projects

## 9.8.1 Timing of events

Figure 9.12 shows a network, with a duration added under each activity.

If we take a notional start time of zero for the project, we can find the earliest possible time for each event. The earliest time for event 1 is clearly 0. The earliest time for event 2 is when A finishes, which is three weeks after its earliest start at 0: the earliest time for event 4 is the time when C finishes, which is two weeks after its earliest start at 3 (i.e. week 5). Similarly, the earliest time for event 5 is $4 + 3 = 7$, for event 3 is 2, and for event 7 is $2 + 3 = 5$.

When several activities have to finish before an event, the earliest time for the event is the earliest time by which all preceding activities can finish. The earliest time for event 6 is when both E and F finish. E can finish one week after its earliest start at 5 (i.e. week 6); F can finish three weeks after its earliest start at 7 (i.e. week 10). Then the

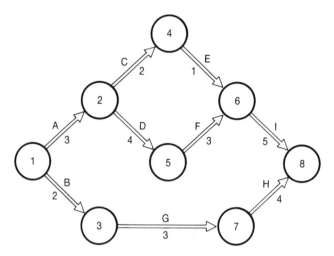

**Figure 9.12**   Network for event timing.

earliest time when both of these can finish is week 10. Similarly, event 8 must wait until both activities H and I finish. Activity H can finish by week $5 + 4 = 9$, while activity I can finish by week $10 + 5 = 15$. The earliest time for event 8 is the later of these which week 15. This gives the overall duration of the project as fifteen weeks. Figure 9.13 shows these earliest event times added to the network.

Now we can do a similar analysis to find the latest time for each event. For this we start at the end of the project with event 8, which has a latest time of week 15. To

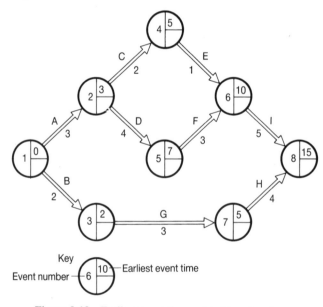

**Figure 9.13**   Earliest event times added to network.

allow activity I to finish by week 15 it must start five weeks before this, so the latest time for event 6 is week $15 - 5 = 10$. The latest H can finish is week 15, so the latest time it can start is four weeks before this, so the latest time for event 7 is week $15 - 4 = 11$. Similarly the latest time for event 3 is $11 - 3 = 8$, for event 5 is $10 - 3 = 7$, and for event 4 is $10 - 1 = 9$.

For events that have more than one following activity, the latest time must allow all following activities to finish on time. Event 2 is followed by activities C and D; C must finish by week 9 so it must start two weeks before this (i.e. week 7), while D must finish by week 7 so it must start four weeks before this (i.e. week 3). The latest time for event 2 which allows both C and D to start on time is the earlier of these, which is week 3.

Similarly, the latest time for event 1 must allow both A and B to finish on time. The latest start time for B is $8 - 2 = 6$ and the latest start time for A is $3 - 3 = 0$. The latest time for event 1 must allow both of these to start on time and this means a latest time of 0. Figure 9.14 shows the network with latest times added for each event.

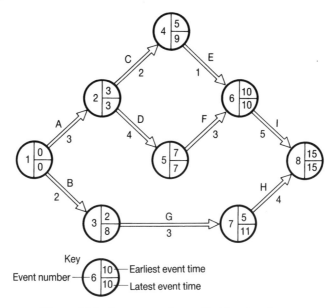

**Figure 9.14**   Laest event times added to network.

## In summary

**Finding the timing of events is an important part of project planning. We can find an earliest and a latest time for each event.**

## 9.8.2 Timing of activities

The earliest start time for an activity is the earliest time of the preceding event: the earliest finish time is the earliest start time plus the duration (shown in Figure 9.15).

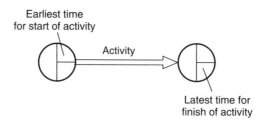

Earliest time
for start of activity

Activity

Latest time for
finish of activity

**Figure 9.15**   Earliest and latest activity times.

Looking at one activity in Figure 9.14, say G, the earliest start time is week 2 and the earliest finish time is, therefore, week $2 + 3 = 5$.

The latest finish time for each activity is the latest time of the following event: the latest start time is the latest finish time minus the duration. For activity G the latest finish is week 11 and the latest start is week $11 - 3 = 8$.

Repeating these calculations for all activities in the project gives the results in Figure 9.16.

| | A | B | C | D | E | F | G |
|---|---|---|---|---|---|---|---|
| 1 | **Project timing** | | | | | | |
| 2 | | | | | | | |
| 3 | **Activity** | **Duration** | **Earliest start** | **Earliest finish** | **Latest start** | **Latest finish** | **Float** |
| 4 | A | 3 | 0 | 3 | 0 | 3 | 0 |
| 5 | B | 2 | 0 | 2 | 6 | 8 | 6 |
| 6 | C | 2 | 3 | 5 | 7 | 9 | 4 |
| 7 | D | 4 | 3 | 7 | 3 | 7 | 0 |
| 8 | E | 1 | 5 | 6 | 9 | 10 | 4 |
| 9 | F | 3 | 7 | 10 | 7 | 10 | 0 |
| 10 | G | 3 | 2 | 5 | 8 | 11 | 6 |
| 11 | H | 4 | 5 | 9 | 11 | 15 | 6 |
| 12 | I | 5 | 10 | 15 | 10 | 15 | 0 |

**Figure 9.16**   Timing of activities in projects.

As you can see from Figure 9.16, some activities have flexibility in time: activity G can start as early as week 2 or as late as week 8, while activity C can start as early as week 3 or as late as week 7. Some activities though, have no flexibility at all: activities A, D, F, and I are at fixed times as their latest start time is the same as their earliest start time. The **total float** measures the amount of flexibility in timing.

**total float** = latest finish − earliest start − duration

This gives the difference between the time available for an activity and the time actually used – so it measures the amount that an activity can expand without affecting the duration of the project. Calculating the total float for activity G in the example has:

Latest finish = latest time of following event (7) = 11

Earliest start = earliest time of preceding event (3) = 2

Duration of activity G = 3

So total float = latest finish – earliest start – duration = $11 - 2 - 3 = 6$

This means that G can expand by up to six weeks without affecting the duration of the project.

The activities that have to be done at a fixed time are called the **critical activities**, and they form a continuous path through the network, called the **critical path**. The length of this path sets the overall project duration. If one of the critical activities is extended or delayed by a certain amount, the overall project duration is also extended by this amount. The activities that have some flexibility in timing are the **non-critical activities** and these may be delayed or extended without necessarily affecting the project duration.

## we Worked example 9.10

A small telephone exchange is planned with ten main activities. Durations (in weeks) and dependencies are shown Table 9.9.

Draw the network for this project, find its duration and calculate the total float of each activity.

**Table 9.9**

| Activity | Description | Duration | Depends on |
|---|---|---|---|
| A | Design internal equipment | 10 | – |
| B | Design exchange building | 5 | A |
| C | Order parts for equipment | 3 | A |
| D | Order material for building | 2 | B |
| E | Wait for equipment parts | 15 | C |
| F | Wait for building material | 10 | D |
| G | Employ equipment assemblers | 5 | A |
| H | Employ building workers | 4 | B |
| I | Install equipment | 20 | E,G,J |
| J | Complete building | 30 | F,H |

What happens if activity A is delayed by six weeks? What happens if activity C is delayed by six weeks?

## Solution

The network for this is shown in Figure 9.17 with timing, including float, calculated in Figure 9.18.

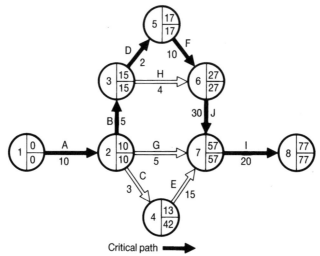

Critical path ➤

**Figure 9.17**  Network for Worked Example 9.10.

|   | A | B | C | D | E | F | G |
|---|---|---|---|---|---|---|---|
| 1 | **Project timing** | | | | | | |
| 2 | | | | | | | |
| 3 | **Activity** | **Duration** | **Earliest start** | **Earliest finish** | **Latest start** | **Latest finish** | **Float** |
| 4 | A | 10 | 0 | 10 | 0 | 10 | 0 |
| 5 | B | 5 | 10 | 15 | 10 | 15 | 0 |
| 6 | C | 3 | 10 | 13 | 39 | 42 | 29 |
| 7 | D | 2 | 15 | 17 | 15 | 17 | 0 |
| 8 | E | 15 | 13 | 28 | 42 | 57 | 29 |
| 9 | F | 10 | 17 | 27 | 17 | 27 | 0 |
| 10 | G | 5 | 10 | 15 | 52 | 57 | 42 |
| 11 | H | 4 | 15 | 19 | 23 | 27 | 8 |
| 12 | I | 20 | 57 | 77 | 57 | 77 | 0 |
| 13 | J | 30 | 27 | 57 | 27 | 57 | 0 |

**Figure 9.18**  Timing of project in Worked Example 9.10.

The duration of the project is 77 days, defined by the critical path A, B, D, F, I and J.

Activity A is on the critical path, so if it is delayed by six weeks, the project will be extended by six weeks, to 83 weeks. Activity C is not on the critical path and has a total float of 29 weeks – so a delay of six weeks will have no effect on the project duration.

## In summary

**We can find an earliest and latest start and finish time for each activity. The total float measures the amount of flexibility in these times. Critical activities have no float; they form the critical path which sets the overall project duration.**

## Chapter review

This chapter described some models that are widely used in business. We could have described many areas, but have emphasised problems with inventory control, linear programming and network analysis. In particular, the chapter:

- discussed the purpose of stocks;
- described the economic order quantity and related calculations;
- talked about constrained optimisation and linear programmes;
- formulated problems as linear programmes;
- analysed computer solutions to linear programmes;
- described projects and their activities;
- drew a project as a network of alternating activities and events;
- calculated the timing of events and activities.

## Problems

9.1    The demand for an item is constant at 100 units a year. Unit cost is £50, reorder cost is £20 and holding cost is £10 per unit a year. What are the economic order quantity, cycle length and costs?

9.2    A company meets the demand for an item of 100 units a week during a 50-week year. The cost of each unit is £20 and the company aims for a return of 20% on capital invested. Annual warehouse costs are 5% of the value of goods stored. The purchasing department of the company costs £45 000 a year and sends out an average of 2000 orders. Find the optimal order quantity for the item, the optimal time between orders and the minimum cost of stocking the item.

9.3    Demand for an item is steady at 20 units a week and the economic order quantity is 50 units. What is the reorder level when the lead-time is (a) 1 week (b) 2 weeks?

9.4.    Two additives $X_1$ and $X_2$ can be used to increase the octane number of petrol. One kilogramme of $X_1$ in 5000 litres of petrol will increase the octane number by 10, and one kilogramme of $X_2$ in 5000 litres will increase the octane number by 20. The total additives must increase the octane number by at least 5, but a total of no more than half a kilogramme can be added to 5000 litres and the amount of $X_2$ plus twice the amount of $X_1$ must be at least half a kilogramme. If $X_1$ costs £30 a kilogramme and $X_2$ costs £40 a kilogramme, formulate this problem as a linear programme.

9.5    A local authority is planning a number of blocks of flats. Five types of block
       have been designed, containing flats of four categories (senior citizens, single
       person, small family and large family). Some information on these flats is given
       in Table 9.10. The council must build a total of 500 flats with at least 40 in
       category 1 and 125 in each of the other categories. High-rise flats are not popular
       and the council wants to limit the number of storeys in the development. In
       particular, the average number of storeys must be at most 5 and at least half the
       flats must be in blocks of 3 or fewer storeys. An area of 300 units has been set
       aside for the development and any spare land will be used as a park. Formulate
       this problem as a linear programme.

**Table 9.10**

| Type of block | No. of flats of each type | | | | No. of storeys | Plan area | Cost per block (£000) |
|---|---|---|---|---|---|---|---|
| | 1 | 2 | 3 | 4 | | | |
| A | 1 | 2 | 4 | 0 | 3 | 5 | 208 |
| B | 0 | 3 | 6 | 0 | 6 | 5 | 320 |
| C | 2 | 2 | 2 | 4 | 2 | 8 | 300 |
| D | 0 | 6 | 0 | 8 | 8 | 6 | 480 |
| E | 0 | 0 | 10 | 5 | 3 | 4 | 480 |

9.6    An oil company makes two blends of fuel by mixing three oils. The costs and
       daily availability of the oils are given in Table 9.11 and the requirements of the
       blends of fuel in Table 9.12

**Table 9.11**

| Oil | Cost (£/litre) | Amount available (litres) |
|---|---|---|
| A | 0.33 | 5 000 |
| B | 0.40 | 10 000 |
| C | 0.48 | 15 000 |

**Table 9.12**

| | |
|---|---|
| Blend 1 | at least 30% of A |
| | at most 45% of B |
| | at least 250% of C |
| Blend 2 | at most 35% of A |
| | at least 30% of B |
| | at most 40% of C |

Each litre of blend 1 can be sold for £1.00 and each litre of blend 2 can be sold
for £1.20. Long-term contracts need at least 10 000 litres of each blend.
Formulate this problem as a linear programme. Figure 9.19 shows a solution to
this problem. What does it show?

```
                        -=*=-  PROBLEM DATA   -=*=-
```

MAXIMISE (1)   0.67 A1 + 0.87 A2 + 0.6 B1 + 0.8 B2 + 0.52 C1 + 0.72 C2
SUBJECT TO
     (2)   A1 + A2 <= 5000
     (3)   B1 + B2 <= 10000
     (4)   C1 + C2 <= 15000
     (5)   A1 + B1 + C1 <= 10000
     (6)   A2 + B2 + C2 <= 10000
     (7)   0.7 A1 − 0.3 B1 − 0.3 C1 >= 0
     (8)   −0.45 A1 + 0.55 B1 − 0.45 C1 <= 0
     (9)   −0.25 A1 − 0.25 B1 + 0.75 C1 >= 0
    (10)   0.65 A2 - 0.35 B2 − 0.35 C2 <= 0
    (11)   −0.3 A2 + 0.7 B2 − 0.3 C2 >= 0
    (12)   −0.4 A2 - 0.4 B2 + 0.6 C2 <= 0
END OF PROBLEM

```
                        -=*=-  SOLUTION   -=*=-
```

LP OPTIMUM FOUND AT STEP     10
OBJECTIVE FUNCTION VALUE:     21150.00

| VARIABLE | VALUE | SENSITIVITY |
|---|---|---|
| A1 | 3000.00 | 0.00 |
| A2 | 2000.00 | 0.00 |
| B1 | 0.00 | 0.67 |
| B2 | 10000.00 | 0.00 |
| C1 | 7000.00 | 0.00 |
| C2 | 8000.00 | 0.00 |

| ROW | SLACK OR SURPLUS | SHADOW PRICES |
|---|---|---|
| 2 | 0.00 | 1.14 |
| 3 | 0.00 | 1.07 |
| 4 | 0.00 | 0.32 |
| 5 | 0.00 | 0.00 |
| 6 | 100000.00 | 0.00 |
| 7 | 0.00 | −0.67 |
| 8 | 4500.00 | 0.00 |
| 9 | 4500.00 | 0.00 |
| 10 | 5000.00 | 0.00 |
| 11 | 4000.00 | 0.00 |
| 12 | 0.00 | 0.67 |

NUMBER OF ITERATIONS = 10

```
                   -=*=-  SENSITIVITY ANALYSIS   -=*=-
```

OBJ COEFFICIENT RANGES

| VARIABLE | CURRENT COEF | ALLOWABLE INCREASE | ALLOWABLE DECREASE |
|---|---|---|---|
| A1 | 0.67 | 0.33 | 0.53 |
| A2 | 0.87 | 0.53 | 0.33 |
| B1 | 0.60 | 0.67 | INFINITY |
| B2 | 0.80 | INFINITY | 0.67 |
| C1 | 0.52 | 0.29 | 0.23 |
| C2 | 0.72 | 0.80 | 0.29 |

**Figure 9.19**   Printout of solution for Problem 9.6.     (*continued*)

RIGHT-HAND SIDE RANGES

| ROW | CURRENT RHS | ALLOWABLE INCREASE | ALLOWABLE DECREASE |
|-----|-------------|--------------------|--------------------|
| 2 | 5000.00 | 0.00 | 1428.57 |
| 3 | 10000.00 | 0.00 | 4285.71 |
| 4 | 15000.00 | 3333.33 | 0.00 |
| 5 | 10000.00 | 0.00 | INFINITY |
| 6 | 10000.00 | 10000.00 | INFINITY |
| 7 | 0.00 | 1000.00 | 0.00 |
| 8 | 0.00 | INFINITY | 4500.00 |
| 9 | 0.00 | 4500.00 | INFINITY |
| 10 | 0.00 | INFINITY | 5000.00 |
| 11 | 0.00 | 4000.00 | INFINITY |
| 12 | 0.00 | 0.00 | 2000.00 |

END OF SOLUTION

**Figure 9.19 cont'd**    Printout of solution for Problem 9.6.

9.7    A project has the activities described in the dependency table of Table 9.13. Draw the network for this project.

**Table 9.13**

| Activity | Depends on | Activity | Depends on |
|----------|-----------|----------|-----------|
| A | – | G | B |
| B | – | H | G |
| C | A | I | E, F |
| D | A | J | H, I |
| E | C | K | E, F |
| F | B, D | L | K |

9.8    Draw a network for the dependency table of Table 9.14.

**Table 9.14**

| Activity | Depends on | Activity | Depends on |
|----------|-----------|----------|-----------|
| A | H | I | F |
| B | H | J | I |
| C | K | K | L |
| D | I, M, N | L | F |
| E | F | M | O |
| F | – | N | H |
| G | E, L | O | A, B |
| H | E | P | N |

9.9    If each activity in Problem 9.8 has a duration of 1 week, find the earliest and latest times for each event. What are the earliest and latest start and finish times for each activity and the total floats?

9.10 A project consists of ten activities with estimated durations (in weeks) and dependencies shown in Table 9.15.

(a) What are the duration of the project and the earliest and latest times for activities?

**Table 9.15**

| Activity | Depends on | Duration | Activity | Depends on | Duration |
|----------|-----------|----------|----------|-----------|----------|
| A | – | 8 | F | C, D | 10 |
| B | A | 6 | G | B, E, F | 5 |
| C | – | 10 | H | F | 8 |
| D | – | 6 | I | G, H, J | 6 |
| E | C | 2 | J | A | 4 |

(b) If activity B needs special equipment, when should this equipment be scheduled?

(c) A check on the project at week 12 shows that activity F is running two weeks late, that activity J will now take six weeks, and that the equipment for B will not arrive until week 18. What effect does this have on the overall project duration?

## CS Case study – Elemental Electronics

Elemental Electronics buys components from a number of suppliers and assembles microcomputers. It does virtually no research, and is happy to use well tried designs that have been tested by other manufacturers. It also spends little on advertising, preferring to sell computers through a few specialised retailers. As a result, it has very low overheads, and can sell its machines at a much lower price than major competitors.

Elemental makes four models of computer. Each of these has four stages in manufacturing: subassembly, main assembly, final assembly and finishing. These operations use semi-automatic machines, with the times needed and available given in Table 9.16.

**Table 9.16**

| | Hours needed per unit | | | | Number of machines | Hours available per machine per week |
|---|---|---|---|---|---|---|
| | Model A | Model B | Model C | Model D | | |
| Subassembly | 2 | 3 | 4 | 4 | 10 | 40 |
| Main assembly | 1 | 2 | 2 | 3 | 6 | 36 |
| Final assembly | 3 | 3 | 2 | 4 | 12 | 38 |
| Finishing | 2 | 3 | 3 | 3 | 8 | 40 |

Fixed costs of production are £3 million a year. Selling prices are £2500, £2800, £3400 and £4000, respectively, for the four models, with direct costs of £1600, £1800, £2200 and £2500 respectively. On average, the subassembly machines need 10% of their total time for maintenance, main assembly machines need 16.667%, final assembly machines 25% and finishing machines 10%. The company works a standard 48 hour week. What is the best mix of models for Elemental?

part

4

# *Business statistics*

We have now looked, in Parts One to Three, at the background to quantitative methods, how to collect and describe data, and how to solve certain types of problem. This last part describes some useful statistics.

There are four chapters in part Four. This first, Chapter 10, introduces the idea of uncertainty and shows how to measure it using probabilities. Chapters 11–13 describe some models for dealing with uncertainty in business.

# 10 Uncertainty and probabilities

## Contents

## Chapter outline

This chapter looks at uncertainty and how we can measure it with probabilities. After reading the chapter you should be able to:

- discuss problems of uncertainty
- work with probabilities
- calculate the probability of independent events
- understand dependent events
- use Bayes' theorem to calculate conditional probabilities

# 10.1 Measuring uncertainty

## 10.1.1 Introduction

So far we have assumed that variables have fixed values. We could look at sales and say, 'a company sells 1000 units a year at £20 a unit'. Or we could look at employment and say, '60 people work 40 hours a week and earn £10 an hour'. These figures have fixed values, and there is no uncertainty. This kind of situation is called **deterministic**.

In reality, we often cannot be so certain. A company does not know exactly how many sales it will have next year; a manufacturer does not know exactly how many units it will make next year; investors do not know exactly what returns they will get. These situations have uncertainty, and are called **stochastic** or **probabilistic**.

Although stochastic problems have some uncertainty, this is not the same as ignorance. When we spin a coin there is uncertainty about the outcome, but we know that it will come down either heads or tails, and that these are equally likely. When a company launches a new product, it may not know its future sales exactly, but it will do enough market research to have some idea. Therefore uncertainty means that although we do not know exactly what is going to happen, we do have some information about the likely outcomes.

In this chapter we are going to measure uncertainty – and we do this with **probabilities**.

> The **probability** of an event gives its likelihood or relative frequency.

You know that when you toss a fair coin there are two equally likely outcomes – heads and tails. The coin comes down heads half the time, so we can say that, the probability of a fair coin coming down heads is 0.5.

$$\text{probability of an event} = \frac{\text{number of times the event occurs}}{\text{total number of observations}}$$

There are 52 cards in a pack and one is the ace of hearts – so choosing a card at random gives a probability of 1/52 that it is the ace of hearts. In the last 500 days the train to work has broken down 10 times, so the probability of it breaking down on any day is 10/500 or 0.02. For 200 of the last 240 trading days the Dow Jones Index has advanced, so there is a probability of 200/240 or 0.83 that the index advanced on a particular day.

As the probability measures the *proportion* of times that an event occurs, it is defined in the range 0 to 1.

- Probability = 0 means the event will **never** occur
- Probability = 1 means the event will **always** occur
- Probability between 0 and 1 gives relative frequency
- Probability outside the range 0 to 1 has no meaning

An event with a probability of 0.8 is quite likely – as it will happen eight times out of ten; an event with a probability of 0.5 is as likely to happen as not; an event with a probability of 0.2 is quite unlikely – as it will happen two times out of ten.

We can abbreviate 'the probability of an event is 0.8' to $P(\text{event}) = 0.8$. Then spinning a coin has $P(\text{head}) = 0.5$.

# we Worked example 10.1

A magazine runs a prize draw with 1 first prize, 5 second prizes, 100 third prizes and 1000 fourth prizes. Prizewinners are drawn at random from entries, and after each draw the winning entry is returned to the draw. By the closing date there are 10 000 entries. If no entry won more than one prize, what is the probability that a given entry won first prize or that it won any prize?

## Solution

● There are 10 000 entries and 1 first prize, so the probability of an entry winning first prize is 1/10,000.

● There are 5 second prizes, so the probability of an entry winning one of these is 5/10 000. The probabilities of winning third or fourth prizes are 100/10 000 and 1000/10 000 respectively.

● In all there are 1106 prizes, so the probability of an entry winning one of these is 1106/10 000 = 0.1106. The probability of not winning a prize is 8894/10 000 = 0.8894.

# we Worked example 10.2

An office has the following types of employee:

|  | Female | Male |
| --- | --- | --- |
| Administrative | 25 | 15 |
| Operational | 35 | 25 |

If one person from the office is chosen at random, what is the probability the person is (a) a male administrator, (b) a female operator, (c) male, (d) an operator?

## Solution

(a)  There are 100 people working in the office. Of these, 15 are male administrators, so:

$$P(\text{male administrator}) = 15/100 = 0.15$$

(b)  35 people in the office are female operators, so:

$$P(\text{female operator}) = 35/100 = 0.35$$

(c)  40 people in the office are male, so:

$$P(\text{male}) = 40/100 = 0.4$$

(d)  60 people in the office are operators, so:

$$P(\text{operator}) = 60/100 = 0.6$$

These worked examples illustrate the two ways of finding probabilities.

- Use theoretical argument to give *a priori* probabilities.

$$\text{probability of an event} = \frac{\text{number of ways the event can occur}}{\text{number of possible outcomes}}$$

The probability that a husband and wife share the same birthday is 1/365 (ignoring leap years). This is an *a priori* probability found by saying that there are 365 days on which the second partner can have a birthday, and only one of these is the same as the first partner's.

- We can use historical data to give **empirical** probabilities.

$$\text{probability of an event} = \frac{\text{number of times the event occurred}}{\text{number of observations}}$$

There was a crowd of more than 20 000 in 62 of the last 100 matches that a football team played at home. This gives an empirical probability of $62/100 = 0.62$ that next week's game will have a crowd of more than 20 000.

You can only get reliable empirical values by collecting typical values over a reasonably long time. Small amounts of historical data may not be typical. You can, for example toss a coin five times and get heads each time, but this does not mean that the coin will always come down heads.

There is one other method of getting probabilities, but it should be avoided. This asks people to give their subjective views about probabilities. You might, for example, ask a financial expert for a probability that a company will make a profit next year. This is equivalent to judgemental forecasting, and has the same drawbacks of unreliability.

## In summary

**Many situations contain some uncertainty. Probabilities give a way of measuring this, by describing the likelihood that an event will happen.**

### 10.1.2 Calculations with probabilities

An important idea when working with probabilities is **mutually exclusive** events. Two events are mutually exclusive if one event happening means the second event cannot happen. When you toss a coin, having it come down heads is mutually exclusive with having it come down tails; a company making a profit one year is mutually exclusive with it making a loss; increasing sales in a month are mutually exclusive with decreasing sales, and so on.

We can do some calculations for the probability of mutually exclusive events. In particular, we can find the probability that one event **or** another happens by adding the separate probabilities.

For **mutually exclusive** events:

OR means ADD probabilities

$P(a \text{ OR } b) = P(a) + P(b)$

$P(a \text{ OR } b \text{ OR } c) = P(a) + P(b) + P(c)$

etc.

We saw this idea in Worked Example 10.1, when we found that the probability that an entry won a prize draw was 1106/10000, and the probability that it did not win was 8894/10000. Each ticket must either win or lose, and these two events are mutually exclusive, so:

$$P(\text{win OR lose}) = P(\text{win}) + P(\text{lose}) = 0.1106 + 0.8894 = 1.0$$

## we Worked example 10.3

A company makes 40 000 washing machines a year. Of these 10,000 are for the home market, 8000 are exported to North America, 7000 to Europe, 5000 to South America, 4000 to the Far East, 3000 to Australasia and 3000 to other markets.

(a)  What is the probability that a particular machine is sold on the home market?

(b)  What is the probability that a machine is exported?

(c)  What is the probability that a machine is exported to either North or South America?

(d)  What is the probability that a machine is sold in either the home market or Europe?

## Solution

(a) The probability that a machine is sold on the home market is:

$$P(\text{home}) = \frac{\text{number sold on home market}}{\text{number sold}} = \frac{10\,000}{40\,000} = 0.25$$

(b) The fact that a machine is sold on the home market is mutually exclusive with the fact that it is exported. Since all machines made are sold somewhere:

$$P(\text{sold}) = 1 = P(\text{exported}) + P(\text{home})$$

So:

$$P(\text{exported}) = 1 - P(\text{home}) = 1 - 0.25 = 0.75$$

(c) $P(\text{North America OR South America}) = P(\text{North America}) + P(\text{South America})$

$$= \frac{8000}{40\,000} + \frac{5000}{40\,000} = 0.2 + 0.125 = 0.325$$

(d) $P(\text{home OR Europe}) = P(\text{home}) + P(\text{Europe})$

$$= \frac{10\,000}{40\,000} + \frac{7000}{40\,000} = 0.25 + 0.175 = 0.425$$

**Venn diagrams** give a useful way of presenting probabilities. These show the probabilities of events by means of circles. If two events are mutually exclusive, then the Venn diagram shows separate circles, as in Figure 10.1.

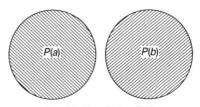

$$P(a \text{ OR } b) = P(a) + P(b)$$

**Figure 10.1**    Venn diagram for mutually exclusive events.

If two events are not mutually exclusive, there is a probability they can both happen. This is shown in the Venn diagram of Figure 10.2. The circles now overlap, with the overlap showing the probability that both events happen. Adding the probabilities of events that are not mutually exclusive would add this overlap twice. To correct for this we have to subtract the probability of both events.

> For events that are *not* mutually exclusive:
>
> P(a OR b) = P(a) + P(b) − P(a AND b)

If you pick a single card from a pack, the probability that it is an ace is 4/52, and the probability that it is a heart is 13/52. But these events are not mutually exclusive and:

$$P(\text{ace OR heart}) = P(\text{ace}) + P(\text{heart}) - P(\text{ace AND heart})$$

$$= \frac{4}{52} + \frac{13}{52} - \frac{1}{52} = \frac{16}{52} = 0.31$$

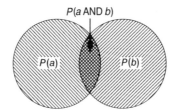

P(a OR b) = P(a) + P(b) − P(a AND b)

**Figure 10.2**   Venn diagram for non-mutually exclusive events.

Another important idea for probabilities is independence. If the occurrence of one event does not affect the occurrence of a second event, the two events are said to be **independent**. The fact that a person works in a bank is independent of the fact that they are left-handed. The fact that a factory has a shipment of raw materials delayed is independent of the fact that it has a problem with machine reliability. Using the notation:

$P(a) = $ the probability of event $a$

$P(a/b) = $ the probability of event $a$ given that event $b$ has already happened

$P(a/\underline{b}) = $ the probability of event $a$ given that $b$ has *not* happened

the two events $a$ and $b$ are independent if:

$$P(a) = P(a/b) = P(a/\underline{b})$$

The probability that a person buys a particular newspaper is independent of the probability that they suffer from hayfever. Then:

$$P(\text{buys } Times) = P(\text{buys } Times/\text{suffers from hayfever})$$
$$= P(\text{buys } Times/\text{does not suffer from hay fever})$$

When a number of independent events all happen, the probability of one and another happening comes from multiplying the probabilities of the separate events.

For **indepedent** events:

AND means MULTIPLY probabilities

$P(a \text{ AND } b) = P(a) \times P(b)$

$P(a \text{ AND } b \text{ AND } c) = P(a) \times P(b) \times P(c)$

etc.

## we Worked example 10.4

A workshop combines two parts, $X$ and $Y$, into a final assembly. An average of 10% of $X$ are defective and 5% of $Y$. If defects on $X$ and $Y$ are independent, what is the probability that a final assembly has both $X$ and $Y$ defective?

## Solution

For independent events:

$$P(X \text{ defective AND } Y \text{ defective}) = P(X \text{ defective}) \times P(Y \text{ defective})$$
$$= 0.1 \times 0.05 = 0.005$$

Similarly:

$$P(X \text{ defective AND } Y \text{ not defective}) = 0.1 \times 0.95 = 0.095$$

$$P(X \text{ not defective AND } Y \text{ defective}) = 0.9 \times 0.05 = 0.045$$

$$P(X \text{ not defective AND } Y \text{ not defective}) = 0.9 \times 0.95 = 0.855$$

These are the only four possible combinations, so you can check that the probabilities add to 1.

## we Worked example 10.5

A warehouse classifies its stock into three categories $A$, $B$ and $C$. On all category $A$ items it promises a service level of 97% – so there is a probability of 0.97 that the warehouse can meet demand immediately from stock. On category $B$ and $C$ items it promises service levels of 94% and 90%, respectively. If service levels are independent, what are the probabilities that the warehouse can immediately supply an order for:

(a) one item of category $A$ and one item of category $B$

(b) one item from each category

(c) two items of category $A$ and one $B$ and three from category $C$

## Solution

(a) As the events are independent we can multiply the probabilities to give:

$$P(\text{one } A \text{ AND one } B) = P(\text{one } A) \times P(\text{one } B) = 0.97 \times 0.94 = 0.912$$

(b)
$$P(\text{one } A \text{ AND one } B \text{ AND one } C) = P(\text{one } A) \times P(\text{one } B) \times P(\text{one } C)$$
$$= 0.97 \times 0.94 \times 0.90 = 0.821$$

(c)
$$P(\text{two } A \text{ AND one } B \text{ AND three } C) = P(\text{two } A) \times P(\text{one } B) \times P(\text{three } C)$$

We can break this down a bit more by noticing that the probability of two items of category $A$ is the probability that the first item is there AND the second item is there. In other words

$$P(\text{two } A) = P(\text{one } A \text{ AND one } A) = P(\text{one } A) \times P(\text{one } A) = P(\text{one } A)^2$$

Similarly:

$$P(\text{three } C) = P(\text{one } C)^3$$

Then the answer is $P(\text{one } A)^2 \times P(\text{one } B) \times P(\text{one } C)^3 = 0.97^2 \times 0.94 \times 0.9^3$
$$= 0.645$$

### In summary

**For mutually exclusive events OR means ADD probabilities, so:**

$$P(a \text{ OR } b) = P(a) + P(b)$$

**For independent events AND means MULTIPLY probabilities, so:**

$$P(a \text{ AND } b) = P(a) \times P(b)$$

# 10.2 Conditional probability

## 10.2.1 Bayes' theorem

If events are not independent, then, obviously, they are dependent – and the probability that one happens depends on whether the other happens. The probability that a person is a heart surgeon depends on whether they went to medical school; the probability that a machine will break down depends on its age; the probability that there is a mistake on an invoice depends on the company sending the invoice, and so on.

For dependent events, the fact that one event has occurred – or has not – changes the probability that a second event will occur. Then with:

$P(a)$ = the probability of event $a$

$P(a/b)$ = the probability of event $a$ given that event $b$ has already happened

$P(a/\underline{b})$ = the probability of event $a$ given that $b$ has *not* happened

the two events, $a$ and $b$, are dependent if:

$$P(a) \neq P(a/b) \neq P(a/\underline{b})$$

(Remember that $\neq$ means 'is not equal to'.)

The probability that a company's share price rises depends on whether the company makes a large profit. So:

$$P(\text{share price rises}) \neq P(\text{share price rises/makes a large profit})$$
$$\neq P(\text{share price rises/does not make a large profit})$$

Probabilities in the form $P(a/b)$ are called **conditional** probabilities. There is one important result for conditional probabilities. The probability of two dependent events happening is the probability of the first, multiplied by the conditional probability that the second happens, given that the first has already happened. We can write this rather clumsy statement as:

> For **dependent events**:
>
> $$P(a \text{ AND } b) = P(a) \times P(b/a)$$

With a little thought we can extend this result to:

$$P(a \text{ AND } b) = P(a) \times P(b/a) = P(b) \times P(a/b)$$

Then taking the second two terms and rearranging them gives **Bayes' theorem**.

> **Bayes' theorem**   $P(a/b) = \dfrac{P(b/a) \times P(a)}{P(b)}$

## we Worked example 10.6

Jim Walton is considering an investment. Before he makes a final decision, he can ask for advice from a financial expert. This adviser is not completely reliable, but in the past when an investment has turned out to be good, she has recommended it 75% of times, and when an investment has turned out to be bad, she has not recommended it 80% of times. Jim knows that about 60% of investments are good. How can he describe the reliability of his adviser?

## Solution

Jim knows that the probability that his adviser recommends an investment given that it is good, but he really wants this the other way around – he wants the probability that an investment is good given that his adviser recommends it. If we use the abbreviations AR for adviser recommends, ANR for adviser does not recommend, GI for good investment and PI for poor investment, then Jim knows $P(AR/GI)$ but he wants $P(GI/AR)$. To get from the first of these to the second, Jim has to use Bayes' theorem:

$$P(GI/AR) = \frac{P(AR/GI) \times P(GI)}{P(AR)}$$

To use this equation, Jim needs $P(AR)$ – the probability that his adviser recommends an investment. We can find this from the information we already have. The adviser will either recommend an investment given that it is good, or she will make a mistake and recommend an investment given that it is poor. So:

> the probability that the adviser recommends an investment = to the probability that an investment is good and the advisor recommends it given that it is good, + the probability that an investment is poor and the adviser recommends it given that it is poor.

To put this more briefly:

$$P(AR) = P(AR/GI) \times P(GI) + P(AR/PI) \times P(PI)$$

We can substitute the values we know to get:

$$P(AR) = P(AR/GI) \times P(GI) + P(AR/PI) \times P(PI)$$
$$= (0.75 \times 0.6) + (0.2 \times 0.4) = 0.53$$

Now using Bayes' theorem gives:

$$P(GI/AR) = \frac{P(AR/GI) \times P(GI)}{P(AR)}$$

$$= \frac{0.75 \times 0.6}{0.53} = 0.85$$

This means that if the financial adviser recommends the investment there is a probability of 0.85 that it is a good investment.

The probability the advisor does not recommend the investment is:

$$P(ANR) = 1 - P(AR) = 1 - 0.53 = 0.47$$

Then Bayes' theorem gives:

$$P(PI/ANR) = \frac{P(ANR/PI) \times P(PI)}{P(ANR)}$$

$$= \frac{0.8 \times 0.4}{0.47} = 0.68$$

So when the adviser does not recommend the investment there is a probability of 0.68 that the investment is poor.

## we Worked example 10.7

Two machines make identical parts which are combined on a production line. The older machine makes 40% of the units, of which 85% are of satisfactory quality. The newer machine makes 60% of the units, of which 92% are of satisfactory quality. A random check further down the production line shows an unusual fault, and the machine that made the unit needs adjusting. What is the probability that the older machine made the unit?

## Solution

This problem is shown in Figure 10.3.

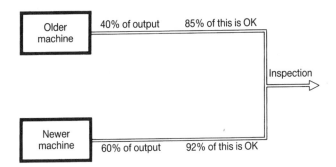

**Figure 10.3** Description of process in Worked Example 10.7.

Using the abbreviations O for the older machine, N for the newer machine, G for good units and F for faulty ones, we want $P(O/F)$ and can find this using Bayes' theorem:

$$P(a/b) = \frac{P(b/a) \times P(a)}{P(b)}$$

so:

$$P(O/F) = \frac{P(F/O) \times P(O)}{P(F)}$$

We know that $P(O) = 0.4$ and $P(F/O) = 0.15$, and we want the value for $P(F)$, the probability that a unit is faulty. You can see that the probability that a unit is faulty is the probability it is faulty from the old machine OR it is faulty from the new machine. So:

$$P(F) = P(F \text{ AND } O) + P(F \text{ AND } N)$$
$$P(F) = P(F/O) \times P(O) + P(F/N) \times P(N)$$
$$= 0.15 \times 0.4 + 0.08 \times 0.6 = 0.108$$

Substituting these values gives:

$$P(O/F) = \frac{P(F/O) \times P(O)}{P(F)} = \frac{0.15 \times 0\ 4}{0.108} = 0.556$$

As always, it is easier to do these calculations with a computer. Figure 10.4 shows a printout of the calculations for this example. The original data are given in the 'prior' and 'conditional' probabilities. The 'marginal' probabilities show the probability that a unit is faulty or good, and the 'revised' probabilities show the results from Bayes' theorem. These figures confirm the value of $P(O/F) = 0.556$, and give the related probabilities for $P(N/F) = 0.444$ and so on.

| | **** DATA ENTERED **** | | |
|---|---|---|---|
| PRIOR PROBABILITIES | P(O) | P(N) | |
| 1 Older machine, P(O) | 0.400 | | |
| 2 Newer machine, P(N) | 0.600 | | |
| CONDITIONAL PROBABILITIES | P(G/O) | P(F/N) | etc. |
| | 1 Faulty, F | 2 Good, G | |
| 1 Older machine, O | 0.150 | 0.850 | |
| 2 Newer machine, N | 0.080 | 0.920 | |
| **** RESULTS CALCULATED **** | | | |
| JOINT PROBABILITIES | P(O and G) | P(N and F) | etc. |
| | 1 Faulty, F | 2 Good, G | |
| 1 Older machine, O | 0.060 | 0.340 | |
| 2 Newer machine, N | 0.048 | 0.552 | |
| MARGINAL PROBABILITIES | P(F) | P(G) | |
| 1 Faulty, P(F) | 0.108 | | |
| 2 Good, P(G) | 0.892 | | |
| RESULTS FOR BAYES' THEOREM | | | |
| Revised probabilities | P(O/G) | P(N/F) | etc. |
| | 1 Faulty, F | 2 Good, G | |
| 1 Older machine, O | 0.556 | 0.381 | |
| 2 Newer machine, N | 0.444 | 0.619 | |

**Figure 10.4**   Printout for Bayes' theorem calculation for Worked Example 10.7.

## In summary

**When two events are dependent we can find conditional probabilities. Bayes' theorem gives a widely used result for conditional probabilities.**

# Chapter review

This chapter introduced the ideas of uncertainty and probability. In particular, it:

- discussed uncertainty in problems;
- defined the probability of an event;
- looked at probabilities of independent events;
- found probabilities for mutually exclusive events;
- discussed conditional probabilities and calculations using Bayes' theorem.

## Problems

10.1 The types of employee in an office are shown in Table 10.1.

**Table 10.1**

|  | Female | Male |
|---|---|---|
| Administrative | 20 | 21 |
| Managerial | 12 | 10 |
| Operational | 42 | 38 |

If one person from the office is chosen at random, what is the probability the person is:

(a) a male administrator

(b) a female manager

(c) male

(d) an operator

(e) either a manager or administrator

(f) either a female administrator or a female manager?

10.2 Four mutually exclusive events A, B, C and D have probabilities of 0.1, 0.2, 0.3 and 0.4 respectively. What are the probabilities of:

(a) A and B

(b) A or B

(c) neither A nor B

(d) A and B and C

(e) A or B or C

(f) none of A, B or C

10.3   If a card is picked from a pack, what is the probability that it is:

(a) an ace

(b) a heart

(c) an ace and a heart

(d) an ace or a heart?

10.4   A salesman schedules three calls for a particular day, and each call has a probability of 0.5 of making a sale. What is the probability of making

(a) 3 sales

(b) 2 or more sales

(c) no sales?

10.5   (a) If $P(a) = 0.4$ and $P(b/a) = 0.3$, what is $P(a \text{ AND } b)$?

(b) If $P(b) = 0.6$, what is $P(a/b)$?

10.6   The probabilities of two events ONE and TWO are 0.4 and 0.6 respectively. The conditional probabilities of three other events A, B and C occurring, given that ONE or TWO has already occurred, are given in Table 10.2.

**Table 10.2**

|      | A   | B   | C   |
|------|-----|-----|-----|
| ONE  | 0.2 | 0.5 | 0.3 |
| TWO  | 0.6 | 0.1 | 0.3 |

What are the conditional probabilities of ONE and TWO occurring given that A, B or C has already occurred?

10.7   A manufacturer uses three suppliers for a component. X supplies 35% of the component, Y supplies 25% and Z supplies the rest. The quality of each component is described as good, acceptable or poor, with the proportions from each supplier given in Table 10.3. What information can you find using Bayes' theorem on these figures?

**Table 10.3**

|   | Good | Acceptable | Poor |
|---|------|------------|------|
| X | 0.2  | 0.7        | 0.1  |
| Y | 0.3  | 0.65       | 0.05 |
| Z | 0.1  | 0.8        | 0.1  |

# CS Case study – The Gamblers' Press

The Gamblers' Press is a weekly paper that publishes information for gamblers. Its main contents are detailed sections on horse racing, greyhound racing, football and other major sporting activities. It also runs regular features on card games, casinos and other gambling events.

The Gamblers' Press is a well respected paper and has a strict policy of giving only factual information. It never gives tips or advice. Last year it decided to run a special feature on misleading or dishonest practices. This idea was suggested when four unconnected reports were passed to the editors.

The first report described an 'infallible' way of winning at roulette. Customers were charged £250 for the details of the scheme, which was based on a record of all the numbers that won during an evening. Then customers were advised to bet on two sets of numbers:

- those that had come up more often, because the wheel might be biased in their favour;

- those that had come up least often, because the laws of probability say that numbers which appear less frequently on one night, must appear more frequently on another night.

The second report showed that a number of illegal chain letters were circulating in the South East. These letters contained a list of eight names. Individuals were asked to send a pound to the name at the top of the list. Then they should delete the name at the top, insert their own name at the bottom, and send a copy of the letter to eight of their friends. As each name moved to the top of the list they would get money from people who joined the chain later. The advertising accompanying these letters guaranteed to make respondents millionaires, and claimed that 'you cannot lose!!!' It also said that people not responding would be letting down their friends and would inevitably be plagued by bad luck.

The third report was from 'a horse racing consultant'. This person sent a letter saying which horse would win a race the following week. A week later he sent a second letter saying the chosen horse had won, and giving another tip for the following week. This was repeated for a third week. Then, after three wins, the consultant said he would send the name of a fourth horse that was guaranteed to win next week, but this time there would be a cost of £1000. This seemed a reasonable price as the horse was certain to win. Unfortunately, this scheme had a drawback. It was thought that the consultant sent out 10000 copies of the original letter, and randomly tipped each horse in a five horse race. The second letter was sent only to those people who had been given the winning horse. The next two letters followed the same pattern, with follow-up letters only sent to those who had been given the winning horse.

The fourth report concerned a North American lottery. Here people chose six numbers in the range 00 to 99 and bought a lottery ticket for $1 containing these numbers. At the end of a week a computer randomly generated a set of six numbers.

Anyone with the same six numbers would win the major prize (typically several million dollars), and people with four or five matching numbers would win smaller prizes. A magazine reported a way of dramatically increasing the chances of winning. This took your eight favourite numbers and bet on all possible combinations of six numbers from these eight. The advertisement explained the benefit of this by saying;

> 'Suppose there is a chance of one in a million of winning the first prize. If one of your lucky numbers is chosen by the computer, you will have this number in over a hundred entries, so you chances of winning are raised by 100 to only 1 in 10 000.'

The Gamblers' Press knew many schemes like these four, and they decided to write a major article on them. What do you think it could say in this article?

# 11

# Probability distributions

**Contents**

## Chapter outline

The previous chapter introduced the idea of probability. This chapter looks at probability distributions. After reading the chapter you should be able to:

● understand the purpose of probability distributions

● draw empirical probability distributions

● do calculations with combinations and permutations

● use the binomial distribution

● use the Poisson distribution

● use the normal distribution

# 11.1 What are probability distributions?

Chapter 4 showed how frequency distributions give the number of observations in different classes. The previous chapter talked about probabilities. Now we are going to combine these two ideas and describe a set of data by a **probability distribution**.

> A **probability distribution** shows the probability that observations are in different classes.

# we Worked example 11.1

Every month the Elkandra Housing Association has a number of people who do not pay their rent. In a typical period, the number of non-payments was as follows.

| | | | | | | | | | | | | | | | | | |
|---|---|---|---|---|---|---|---|---|---|---|---|---|---|---|---|---|---|
| 2 | 4 | 6 | 7 | 1 | 3 | 3 | 5 | 4 | 1 | 2 | 3 | 4 | 3 | 5 | 6 | 2 | 4 |
| 3 | 2 | 5 | 5 | 0 | 3 | 3 | 2 | 1 | 4 | 4 | 4 | 3 | 1 | 3 | 6 | 3 | 4 |
| 2 | 5 | 3 | 4 | 2 | 5 | 3 | 4 | | | | | | | | | | |

Draw a probability distribution of these data. What is the probability that there are more than four non-payers?

## Solution

Adding the number of months with various numbers of non-payers gives the following frequency distribution.

| Non-payers | 0 | 1 | 2 | 3 | 4 | 5 | 6 | 7 |
|---|---|---|---|---|---|---|---|---|
| Frequency | 1 | 4 | 8 | 12 | 10 | 6 | 3 | 1 |

If we divide each of the frequencies by the total number of observations (45) we get the following relative frequency – or probability – distribution.

| No-shows | 0 | 1 | 2 | 3 | 4 | 5 | 6 | 7 |
|---|---|---|---|---|---|---|---|---|
| Probability | 0.02 | 0.09 | 0.18 | 0.27 | 0.22 | 0.13 | 0.07 | 0.02 |

Figure 11.1 shows a histogram of this distribution. Remember that in histograms the area of each rectangle represents the probability, so the total area under the histogram must equal 1.

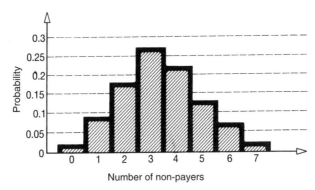

**Figure 11.1**  Probability for non-payers in Worked Example 11.1.

The probability of more than four non-payers is:

$$P(5) + P(6) + P(7) = 0.13 + 0.07 + 0.02 = 0.22$$

The probability distribution in Worked Example 11.1 is empirical – which means that it comes from actual observations. You can draw an empirical probability distribution for any set of observations. Since many empirical distributions follow standard patterns, we can describe some general probability distributions that can be used for many problems. In this chapter we are going to look at the three most important: binomial, poisson and normal distributions.

### In summary

**A probability distribution shows the relative frequency of observations in classes. Empirical distributions describe specific situations, but there are several general distributions.**

# 11.2 Combinations and permutations

The first general distribution is the binomial. For this we need to look at **combinations** and **permutations**.

There are many problems where we want to find the best order for a set of activities. We may, for example, want to find the best order for a bus to visit stops as it travels between two towns; or we may want to find the best order for processing jobs on a machine, or the best order for activities in a project. These problems compare all the possible orders – or sequences – of activities, and find the best.

Sequencing problems seem quite simple, but they are notoriously difficult when the number of possible sequences is very large. If we want to find the best sequence of $n$ activities, we can choose the first as any one of the $n$. Then we can choose the second activity as any one of the remaining $(n - 1)$, so there are $n(n - 1)$ possible sequences for the first two activities. Then the third activity can be any one of the remaining $(n - 2)$, the fourth any of $(n - 3)$, and so on. So the total number of sequences is given by the formula:

$$\text{number of sequences} = n(n - 1)(n - 2)(n - 3)\ldots \times 3 \times 2 \times 1 = n! \quad (n \text{ factorial})$$

Even a small problem with 15 activities has $15! = 1.3 \times 10^{12}$ possible sequences.

Suppose we do not want to find the best sequence of $n$ activities, but want only to choose $r$ of them. In how many ways can we do this? This is where we use permutations and combinations.

The **combination** of $r$ things from $n$ is written as $^nC_r$. This shows the number of ways that we can choose $r$ things from $n$ when the order in which we choose the $r$ things is not important.

> The number of ways that we can choose $r$ things from $n$ when the order of selection is not important is:
>
> $$^nC_r = \frac{n!}{r!(n - r)!}$$

If there is a pool of 10 cars and 3 customers arrive to use them, then there are $^{10}C_3$ ways of assigning customers to cars.

$$^{10}C_3 = \frac{10!}{3! \times (10 - 3)!} = 120$$

If we choose 6 lottery numbers from 49, it does not matter in which order we choose the numbers, so there are $^{49}C_6 = 49!/(6! \times 43!) = 13\,983\,816$ possible choices.

If the order in which we choose things is important, then we want the **permutation** of $r$ things from $n$, which is written as $^nP_r$.

> The number of ways that we can choose $r$ things from $n$ when the order of selection is important is:
>
> $$^nP_r = \frac{n!}{(n - r)!}$$

Suppose there are 10 applicants for a social club's committee consisting of a chairman, deputy chairman, secretary and treasurer. We want to choose a group of 4 from 10, but the order in which we choose them is important as it gives the different jobs. Then the number of ways we can choose a committee of four is:

$$^nP_r = \frac{n!}{(n - r)!} = \frac{10!}{(10 - 4)!} = 5040$$

If four ordinary committee members are chosen – so there is no difference between jobs and the order in which they are chosen is not important – we are interested in the combinations of 4 from 10. The number of ways of choosing ordinary members is:

$$^nC_r = \frac{n!}{r!(n-r)!} = \frac{10!}{4!(10-4)!} = 210$$

Permutations depend on order of selection, and combinations do not, so there are always a lot more permutations than combinations. The number of combinations of four letters of the alphabet is $26!/(4! \times 22!) = 14\,950$. One combination is the letters A, B, C and D, but these can be arranged in 24 different ways (ABCD, ABDC, ACBD, etc.). The number of permutations of four letters from the alphabet is $26!/22! = 358\,800$.

## we Worked example 11.2

Twelve areas in the North Sea are available for oil exploration. The government wants to encourage competition and allocates at most one area to any exploration company.

(a)  If 12 exploration companies bid for the areas, in how many ways can the areas be allocated?

(b)  Initial forecasts show that each area is equally likely to produce oil, so they are equally attractive. If 20 exploration companies put in bids for areas how many ways are there of allocating areas to companies?

(c)  A last minute report shows the probabilities of major oil discoveries in each area. Based on this report, four companies withdraw their bid. In how many ways can the allocation be done now?

## Solution

(a)  This really asks for the number of ways in which the companies can be sequenced. There are 12 companies getting one area each so the companies can be sequenced in 12! ways, or $4.79 \times 10^8$.

(b)  There are 20 companies, only 12 of which are chosen. As each area is equally attractive, the order of selection is not important. Then the number of possible combinations is:

$$^nC_r = \frac{n!}{r!(n-r)!} = \frac{20!}{12!(20-12)!} = 125\,970$$

(c)  Now the areas are different, so the order of selecting the remaining 16 companies is important. This is:

$$^nP_r = \frac{n!}{(n-r)!} = \frac{16!}{(16-12)!} = 8.72 \times 10^{11}$$

## In summary

- There are $n!$ possible sequences of $n$ different things.

- If the order of selection is not important, $r$ things can be chosen from $n$ in $^nC_r$ different ways (these are combinations).

- If the order of selection is important, $r$ things can be chosen from $n$ in $^nP_r$ different ways (these are permutations).

# 11.3 Binomial distribution

The binomial distribution shows the number of successes in a series of 'trials'. Here a trial has the following characteristics:

- each trial is an event with two possible outcomes – success and failure;

- the two outcomes are mutually exclusive;

- there is a constant probability of success, $p$, and failure, $q = 1 - p$;

- the outcomes of successive trials are independent.

Tossing a coin is an example of a binomial process. Each toss is a trial; each head, say, is a success with a constant probability of 0.5, and each tail is a failure. The binomial distribution shows the number of heads to expect in a number of tosses. The binomial distribution can also be used for inspecting products for defects. Each inspection is a trial, each fault is a success and each good unit is a failure. The binomial distribution shows the expected number of faults.

In general, the binomial distribution finds the probability of $r$ successes in $n$ trials, but in each trial the probability of a success is constant at $p$. So:

$$P(\text{exactly } r \text{ successes}) = P(r \text{ successes AND } n - r \text{ failures})$$
$$= P(r \text{ successes}) \times P(n - r \text{ failures})$$

With independent trials, the probability that the first $r$ are successes is $p^r$, and the probability that the next $n - r$ are failures is $q^{n-r}$. Therefore the probability of the first $r$ trials being successes and the next $n - r$ trials being failures is $p^r q^{n-r}$.

However, the sequence of $r$ successes followed by $n - r$ failures is only one way of getting $r$ successes in $n$ trials. In the previous section we saw that the number of sequences of $r$ things chosen from $n$, when the order of selection does not matter, is $^nC_r = n!/r!(n-r)!$ So there must be $^nC_r$ possible sequences of $r$ successes and $n - r$ failures, each with probability $p^r q^{n-r}$. Multiplying the number of sequences by the probability of each gives:

$$P(r \text{ success in } n \text{ trials}) = {}^nC_r\, p^r q^{n-r}$$
$$= \frac{n!}{r!(n-r)!} p^r q^{n-r}$$

## we Worked example 11.3

A salesman has a 50% chance of making a sale when he calls on a customer. One morning he arranges 6 calls.

(a)  What is the probability of his making exactly 3 sales?

(b)  What are the probabilities of making other numbers of sales?

(c)  What is the probability of making fewer than 3 sales?

## Solution

This is a binomial process with $n = 6$ trials. The probability of success (making a sale) is $p = 0.5$, and the probability of failure (not making a sale) is $q = 1 - p = 0.5$.

(a)  The probability of making exactly 3 sales (so $r = 3$) is:

$$P(r \text{ success in } n \text{ trials}) = {}^nC_r \, p^r \, q^{n-r}$$

$$P(3 \text{ success in 6 trials}) = {}^6C_3 \times 0.5^3 \times 0.5^{(6-3)}$$

$$= \frac{6!}{3! \times 3!} \times 0.125 \times 0.125 = 0.3125$$

(b)  Substituting other values for $r$ gives the following values:

| $r$ | 0 | 1 | 2 | 3 | 4 | 5 | 6 |
|---|---|---|---|---|---|---|---|
| $P(r$ successes in 6 trials) | 0.0156 | 0.0938 | 0.2344 | 0.3125 | 0.2344 | 0.0938 | 0.0156 |

These values are shown in the probability distribution in Figure 11.2.

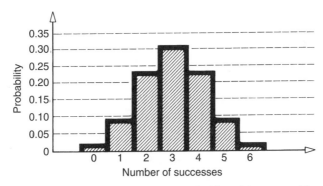

**Figure 11.2**  Probability distribution for binomial process with $n = 6$ and $p = 0.5$.

(c) The probability of making fewer than 3 sales is the sum of the probabilities of making 0, 1 and 2 sales.

$$P(\text{fewer than 3 sales}) = P(0 \text{ sales}) + P(1 \text{ sale}) + P(2 \text{ sales})$$
$$= 0.0156 + 0.0938 + 0.2344$$
$$= 0.3438$$

The shape of the binomial distribution varies with $p$ and $n$. For small values of $p$ the distribution is asymmetrical and the peak is to the left of centre. For larger values of $p$ the peak moves to the centre of the distribution, and with $p = 0.5$ the distribution is symmetrical. For still larger values of $p$ the distribution again becomes asymmetrical but with the peak to the right. For larger values of $n$ the distribution is flatter and broader. Some typical binomial distributions are shown in Figure 11.3.

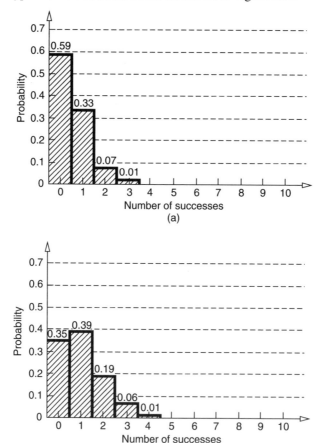

**Figure 11.3** Typical binomial distributions for varying values of $n$ and $p$:
(a) $n = 5$, $p = 0.1$, (b) $n = 10$, $p = 0.1$;

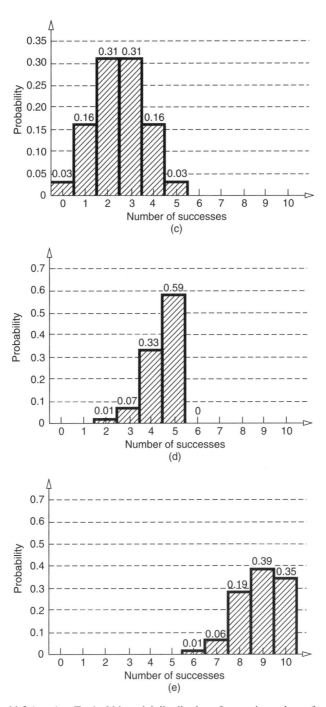

**Figure 11.3 (cont).**   Typical binomial distributions for varying values of $n$ and $p$:
(c) $n = 5$, $p = 0.1$, (d) $n = 5$, $p = 0.9$, (e) $n = 10$, $p = 0.9$.

The mean, variance and standard deviation of a binomial distribution are:

For a binomial distribution:

- mean $= \mu = np$
- variance $= \sigma^2 = npq$
- standard deviation $= \sigma = \sqrt{npq}$

Notice that in these definitions we have used the Greek letter $\mu$ (mu) for the mean rather than $\bar{x}$, and the Greek letter $\sigma$ (sigma) for the standard deviation rather than $s$. This follows a standard notation, where:

- the mean and standard deviation of a sample are called $\bar{x}$ and $s$, while

- the mean and standard deviation of a population are called $\mu$ and $\sigma$.

## we Worked example 11.4

A company makes a sensitive electronic component, and 10% of the output is defective. The components are sold in boxes of 20. Describe the number of defective units in each box.

## Solution

This is a binomial process, with success being a faulty unit. In each box, $n = 20$ and $p = 0.1$. The mean number of faulty units in a box is:

$$np = 20 \times 0.1 = 2$$

The variance is:

$$npq = 20 \times 0.1 \times 0.9 = 1.8$$

and the standard deviation is:

$$npq = \sqrt{1.8} = 1.34$$

## we Worked example 11.5

One evening a market researcher has to visit 12 houses. Previous calls suggest there will be someone at home in 85% of houses.

(a) Describe the probability distribution of the number of houses with people at home.

(b) What is the probability that the researcher will find someone at home in exactly 9 houses?

(c) What is the probability there will be someone at home in exactly 7 houses?

(d) What is the probability there will be someone at home in at least 10 houses?

## Solution

(a) The process is binomial, with visiting a house as a trial and finding someone at home a success. Then we have $n = 12$, $p = 0.85$ and $q = 0.15$, and can substitute these to give:

Mean number of houses with someone at home $= np = 12 \times 0.85 = 10.2$

Variance $= npq = 12 \times 0.85 \times 0.15 = 1.53$

Standard deviation $= \sqrt{1.53} = 1.24$

(b) Let $P(9)$ be the probability that there is someone at home in 9 houses. Then:

$$P(9) = {}^{12}C_9 \times 0.85^9 \times 0.15^3 = 220 \times 0.2316169 \times 0.003375 = 0.172$$

(c) $$P(7) = {}^{12}C_7 \times 0.85^7 \times 0.15^5 = 792 \times 0.320577 \times 0.000076 = 0.0193$$

Rather than do these calculations by hand, you can look them up in the standard tables in Appendix A. These tables only give values for $p$ up to 0.5, so we must redefine success as finding a house with no-one at home. Then $p=0.15$. If you look up the entry for $n = 12$, $p = 0.15$ and $r = 3$ (finding 9 houses with someone at home is the same as finding 3 houses with no-one at home) you can see that the probability is 0.1720, and with $r = 5$ the probability is 0.0193.

(d) Here we want $P(\text{at least } 10)$ which is $P(10) + P(11) + P(12)$

You can find these values in Appendix A:

$$P(\text{at least } 10) = 0.2924 + 0.3012 + 0.1422 = 0.7358$$

You can, of course, use a computer to find binomial probabilities. Figure 11.4 (a) shows typical results from a statistical package giving the binomial probabilities for $n = 20$ and $p = 0.3$. The first command in this Minitab printout asks for a probability distribution function (pdf) and the subcommand describes the distribution. Figure 11.4(b) shows equivalent results from a spreadsheet.

```
MTB  > pdf;
SUBC > binomial n = 20 p = 0.3.

BINOMIAL WITH N = 20 P = 0.300000

     K            P(X = K)
     0            0.0008
     1            0.0068
     2            0.0278
     3            0.0716
     4            0.1304
     5            0.1789
     6            0.1916
     7            0.1643
     8            0.1144
     9            0.0654
    10            0.0308
    11            0.0120
    12            0.0039
    13            0.0010
    14            0.0002
    15            0.0000

MTB > stop
```

(a)

|    | A | B | C |
|----|---|---|---|
| 1  | **Binomial probabilities** | | |
| 2  |   |   |   |
| 3  | n | 20 |   |
| 4  | p | 0.3000 |   |
| 5  |   |   |   |
| 6  | r | **P[r]** | **Cum prob** |
| 7  | 0 | 0.0008 | 0.0008 |
| 8  | 1 | 0.0068 | 0.0076 |
| 9  | 2 | 0.0278 | 0.0355 |
| 10 | 3 | 0.0716 | 0.1071 |
| 11 | 4 | 0.1304 | 0.2375 |
| 12 | 5 | 0.1789 | 0.4164 |
| 13 | 6 | 0.1916 | 0.6080 |
| 14 | 7 | 0.1643 | 0.7723 |
| 15 | 8 | 0.1144 | 0.8867 |
| 16 | 9 | 0.0654 | 0.9520 |
| 17 | 10 | 0.0308 | 0.9829 |
| 18 | 11 | 0.0120 | 0.9949 |
| 19 | 12 | 0.0039 | 0.9987 |
| 20 | 13 | 0.0010 | 0.9997 |
| 21 | 14 | 0.0002 | 1.0000 |
| 22 | 15 | 0.0000 | 1.0000 |

(b)

**Figure 11.4**  Binomial probabilities from (a) a typical statistics package and (b) a spreadsheet.

Sometimes the arithmetic for binomial probabilities gets rather difficult. Suppose the accounts department of a company sends out 10000 invoices a month and, on average, five of these are returned with some error. What is the probability that exactly six invoices will be returned in a month?

Each invoice is a trial, and each error is a success. Then substituting the values given: $n = 10000$, $r = 6$, $p = 5/10000$ and $q = 9995/10000$, we have:

$$P(r \text{ errors}) = \frac{n!}{r!(n-r)!} p^r q^{n-r}$$

$$= \frac{10000!}{6! \times 9994!} \times \left(\frac{5}{10000}\right)^6 \times \left(\frac{9995}{10000}\right)^{9994}$$

It is rather daunting to raise figures to the power of 9994 or to find 10000 factorial. Fortunately, there is an alternative. When $n$ is large and $p$ is small, we can approximate the binomial distribution by a Poisson distribution. This is described in the next section.

## In summary

**We can use the binomial distribution when an event has two mutually exclusive outcomes. It gives the probability of r successes in n trials.**

# 11.4 Poisson distribution

The Poisson distribution is a close relative of the binomial distribution and can be used to approximate it when:

- the number of trials, $n$, is large – greater than about 20;

- the probability of success, $p$, is small – so that $np$ is less than about 5.

The Poisson distribution is also useful in its own right for looking at events that happen at random. The number of accidents each month in a factory, the number of defects in a metre of cloth, and the number of telephone calls received each hour in an office follow Poisson distributions.

The Poisson distribution looks only at the probability of success. It assumes this is very small, so the number of failures is very large – we are looking for a few successes in a continuous background of failures. When we look at the number of spelling mistakes in a long report, we only want the mistakes – or 'successes' – and are not bothered by the words that are spelt properly – the 'failures'.

A Poisson distribution is described by the equation:

$$P(r \text{ successes}) = \frac{e^{-\mu}\mu^r}{r!}$$

where:

e = exponential constant = 2.7183

$\mu$ = mean number of successes

The Poisson distribution can solve the problem that we did not finish above, with the accounts department. Remember that this department sent out 10 000 invoices a month and had an average of five returned with some error. Here $n$ is large, and $np = 5$, so we can use the Poisson distribution to approximate the binomial distribution. We want the probability of exactly six invoices being returned in a month. So $r = 6$ and $\mu = 5$.

$$P(r \text{ successes}) = \frac{e^{-\mu}\mu^r}{r!}$$

so $\quad P(6 \text{ successes}) = \frac{e^{-5}5^6}{6!} = \frac{(0.0067 \times 15\,625)}{720} = 0.1462$

## we Worked example 11.6

A building site had 40 accidents in the past 50 weeks. In what proportion of weeks would it expect 0, 1, 2, 3 and more than 4 accidents?

## Solution

Accidents occur at random over time, so we can use a Poisson distribution. The mean number of accidents a week is $40/50 = 0.8$, so substituting $\mu = 0.8$ and $r = 0$ gives:

$$P(0) = \frac{e^{-0.8} \times 0.8^0}{0!} = 0.4493$$

(Remember that $x^0 = 1$ and $0! = 1$.) Similarly we can substitute $r = 1$ etc. to get:

$$P(1) = \frac{e^{-0.8} \times 0.8^1}{1!} = 0.3595$$

$$P(2) = \frac{e^{-0.8} \times 0.8^2}{2!} = 0.1438$$

$$P(3) = \frac{e^{-0.8} \times 0.8^3}{3!} = 0.0383$$

$$P(4) = \frac{e^{-0.8} \times 0.8^4}{4!} = 0.0077$$

Then:

$$P(>4) = 1 - P(\leq 4)$$

$$= 1 - P(0) - P(1) - P(2) - P(3) - P(4)$$

$$= 1 - 0.4493 - 0.3593 - 0.1438 - 0.0383 - 0.0077$$

$$= 0.0016$$

You can look up Poisson probabilities in the standard tables in Appendix B. Alternatively, you can use a computer. Figure 11.5 shows typical printouts from a statistical package and a spreadsheet.

```
MTB > pdf;
SUBC > poisson mu = 6.

POISSON WITH MEAN = 6.000

     K              P(X = K)
     0              0.0025
     1              0.0149
     2              0.0446
     3              0.0892
     4              0.1339
     5              0.1606
     6              0.1606
     7              0.1377
     8              0.1033
     9              0.0688
    10              0.0413
    11              0.0225
    12              0.0113
    13              0.0052
    14              0.0022
    15              0.0009
    16              0.0003
    17              0.0001
    18              0.0000
MTB > stop
```

(a)

| | A | B | C |
|---|---|---|---|
| 1 | **Poisson probabilities** | | |
| 2 | | | |
| 3 | **mean** | **6** | |
| 4 | | | |
| 5 | **k** | **Prob (k)** | **Cum prob** |
| 6 | 0 | 0.0025 | 0.0025 |
| 7 | 1 | 0.0149 | 0.0174 |
| 8 | 2 | 0.0446 | 0.0620 |
| 9 | 3 | 0.0892 | 0.1512 |
| 10 | 4 | 0.1339 | 0.2851 |
| 11 | 5 | 0.1606 | 0.4457 |
| 12 | 6 | 0.1606 | 0.6063 |
| 13 | 7 | 0.1377 | 0.7440 |
| 14 | 8 | 0.1033 | 0.8472 |
| 15 | 9 | 0.0688 | 0.9161 |
| 16 | 10 | 0.0413 | 0.9574 |
| 17 | 11 | 0.0225 | 0.9799 |
| 18 | 12 | 0.0113 | 0.9912 |
| 19 | 13 | 0.0052 | 0.9964 |
| 20 | 14 | 0.0022 | 0.9986 |
| 21 | 15 | 0.0009 | 0.9995 |
| 22 | 16 | 0.0003 | 0.9998 |
| 23 | 17 | 0.0001 | 0.9999 |
| 24 | 18 | 0.0000 | 1.0000 |

(b)

**Figure 11.5**   Poisson probabilities from (a) a typical statistics package and (b) a spreadsheet.

Random events usually follow a Poisson distribution, but strictly speaking there are some other requirements. In particular, a Poisson process needs:

● events that are independent;

● the probability of an event happening in an interval to be proportional to the length of the interval;

● in theory, an infinite number of events to be possible in an interval.

If these conditions are met, then:

For a **Poisson distribution:**

● mean, $\mu = np$

● variance, $\sigma^2 = np$

● standard deviation, $\sigma = \sqrt{np}$

The shape and position of the Poisson distribution are set by the single parameter $\mu$. For small $\mu$ the distribution is asymmetrical with a peak to the left of centre. For larger values of $\mu$, the distribution is more symmetrical, as shown in Figure 11.6.

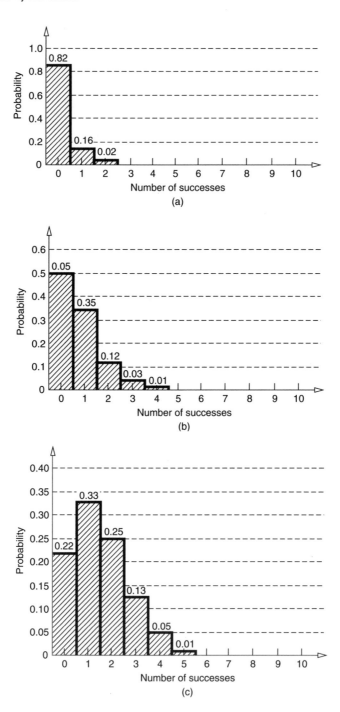

**Figure 11.6**  Typical Poisson distribution for varying values of $\mu$:
(a) $\mu = 0.2$; (b) $\mu = 0.7$, (c) $\mu = 1.5$;

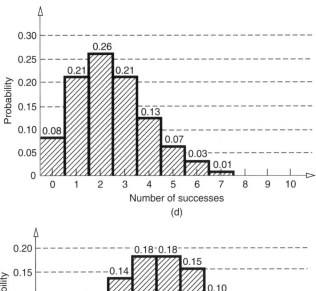

**Figure 11.6 (cont).** Typical Poisson distribution for varying values of $\mu$:
(d) $\mu = 2.5$; (e) $\mu = 5.0$.

## we Worked example 11.7

There is an average of 5 faults in a kilometre of wire.

(a) Describe the distribution of faults.

(b) What is the probability of 2 faults in a kilometre?

### Solution

(a) Random faults follow a Poisson distribution. So the mean number of faults is $\mu = 5$. The variance is also 5, and the standard deviation is $\sqrt{5} = 2.236$.

(b) The probability of $r$ faults is:

$$P(r) = \frac{e^{-\mu}\mu^r}{r!}$$

so

$$P(2) = \frac{e^{-5} \times 5^2}{2!} = 0.0842$$

## we Worked example 11.8

A local council ran a test at a road junction. During this test, cars arrive randomly at the junction at an average rate of 5 cars every 10 minutes.

(a) What is the probability that during a 10-minute period exactly 3 cars arrive?

(b) What is the probability that more than 5 cars arrive in a 10-minute period?

### Solution

(a) Random arrivals follow a Poisson distribution. The mean number of cars arriving in a 10-minute period is $\mu = 5$. Then the probability of exactly 3 cars arriving is:

$$P(r) = \frac{e^{-\mu}\mu^r}{r!}$$

so

$$P(3) = \frac{e^{-5} \times 5^3}{3!} = 0.1404$$

You can check this probability in Appendix B, by looking up the value for $\mu = 5$ and $r = 3$.

(b) The probability that more than 5 cars arrive at the junction in a 10-minute period is:

$$P(>5) = 1 - P(5 \text{ or less})$$
$$= 1 - P(0) - P(1) - P(2) - P(3) - P(4) - P(5)$$
$$= 1 - 0.0067 - 0.0337 - 0.0842 - 0.1404 - 0.1755 - 0.1755$$
$$= 0.384$$

Figure 11.7 shows the probability distribution for this problem.

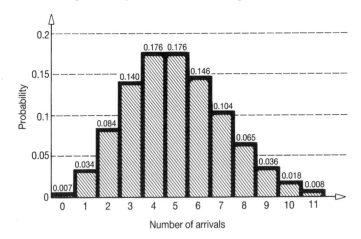

**Figure 11.7** Probability distribution of car arrivals for Worked Example 11.8.

When both $p$ and $r$ are large, the calculations for Poisson distributions become difficult. Suppose a motor insurance policy is only available to certain types of driver. Each of 100 drivers holding the policy has an average of 0.2 accidents each year. What is the probability that fewer than 15 drivers will have accidents in one year?

This is a binomial process with mean $= np = 100 \times 0.2 = 20$. The probability that exactly $r$ drivers will have accidents in the year is:

$$^{20}C_r \times 0.2^r \times 0.8^{100-r}$$

We can find the probability that fewer than 15 drivers will have accidents in the year by adding the results of this calculation for all values of $r$ from 0 to 14. This is rather messy, so we can look for a Poisson approximation. Unfortunately, this approximation only works if $np$ is less than 5, and here it is $100 \times 0.2 = 20$. However, there is another approach, and this time we will use the most common probability distribution of all. When $n$ is large and $np$ is more than 5 the binomial distribution can be approximated by the normal distribution. This is described in the following section.

## In summary

**The Poisson distribution is an approximation to the binomial distribution when the probability of success is small. It can also be used to describe infrequent, random events.**

# 11.5 Normal distribution

The binomial and Poisson distributions both describe discrete data – but the normal distribution describes continuous data. Although these two types of data are similar in principle, there is one important difference. With discrete probabilities we can find the probability of 5 successes in 10 trials, but with continuous data we cannot find the probability that, say, a person weighs 80.456456 kg. The only thing we can do is find the probability that the person weighs between, say, 80.4 and 80.5 kg. Continuous probability distributions give the probability that a value is within a specified range.

The most widely used probability distribution is the **normal distribution**. This is a bellshaped curve which describes many natural phenomena, such as the heights of trees, harvest from an acre of land, weight of horses and daily temperature. It also describes many business functions, such as daily takings in a shop, number of customers a week, size of companies, production in a factory, and so on. The distribution is so common that we can use the 'rule of thumb': for large numbers of observations use the normal distribution.

Figure 11.8 shows a normal distribution. It is the area under this curve – and not its height – that shows the probability.

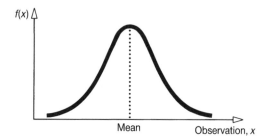

**Figure 11.8** Normal probability distribution.

The **normal distribution** curve:

• is continuous

• is symmetrical about the mean value, $\mu$

• has mean, median and mode all equal

• has total area under the curve equal to 1

• in theory extends to plus and minus infinity on the *x*-axis

Suppose a factory makes boxes of chocolates with a mean weight of 1000 g. There will be small variations in the weight of each box, and if we take a large number of boxes the weights will follow a normal distribution. Managers in the factory will be interested to find, for example, the number of boxes that weigh more than 1005 g. This is given by the area under the right-hand tail of the distribution, as shown in Figure 11.9.

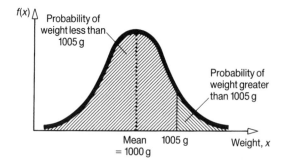

**Figure 11.9** Distribution of weights of chocolate boxes.

There are two ways of finding the area under the tail of the distribution:

● we can use a computer to do the calculation;

● we can look up values in standard tables, shown in Appendix C.

Normal distribution tables are based on a value Z. This is the number of standard deviations that a point is away from the mean. Then normal tables show the probability that a value greater than this will occur.

With the boxes of chocolates, the mean weight is 1000 g and we will assume the standard deviation is 3 g. To find the probability that a box weighs more than 1005 g, we need the area in the tail of the probability distribution, as shown in Figure 11.9. To find this area we calculate Z and look this value up in normal tables, which will give the corresponding probability.

$Z$ = number of standard deviations from the mean

$$= \frac{\text{value} - \text{mean}}{\text{standard deviation}} = \frac{x - \mu}{\sigma}$$

$$= \frac{1005 - 1000}{3}$$

$$= 1.67$$

Looking up 1.67 in the tables in Appendix C gives a value of 0.0475. This is the probability that a box will weigh more than 1005 g.

The normal distribution is symmetrical about the mean, so we can do some other calculations. For example, the probability that a box of chocolates weighs less than 995 g is the same as the probability that it weighs more than 1005 g, which we have already found as 0.0475 (shown in Figure 11.10).

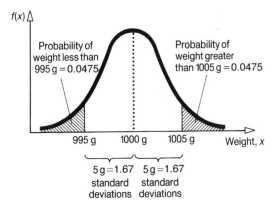

**Figure 11.10**  Symmetrical distribution of weights of chocolate boxes.

## we Worked example 11.9

Figures kept by an auctioneer show that the weight of cattle brought to market in the past five years has a mean of 950 kg and a standard deviation of 150 kg. What proportion of these animals fall into the following categories?

(a) More than 1250 kg.

(b) Less than 850 kg.

(c) Between 1100 and 1250 kg.

(d) Between 800 and 1300 kg.

(e) Below what weight are 95% of cattle?

## Solution

With a large number of cattle brought to market, we can assume a normal distribution with $\mu = 950$ and $\sigma = 150$.

(a) We can find the probability of weight greater than 1250 kg as follows.

$$Z = \text{number of standard deviations from the mean}$$

$$= \frac{1250 - 950}{150} = 2.0$$

Looking this up in the normal tables in Appendix C gives a value of 0.0228. This is the probability we want, as shown in Figure 11.11.

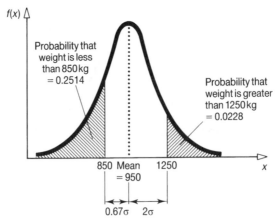

**Figure 11.11** Calculations for Worked Example 11.9.

(b) We can find the probability of weight less than 850 kg in the same way.

$$Z = \frac{850 - 950}{150} = -0.67$$

The table only shows positive values, but as the distribution is symmetrical we can use the value for $+0.67$, which is $0.2514$.

(c) Normal tables only show probabilities in the tail of the distribution, so we often have to do some juggling. To find the probability that the weight is between 1100 kg and 1250 kg, we use:

$P(\text{between } 1100 \text{ and } 1250 \text{ kg}) = P(\text{greater than } 1100 \text{ kg}) - P(\text{greater than } 1250 \text{ kg})$

This is shown in Figure 11.12.

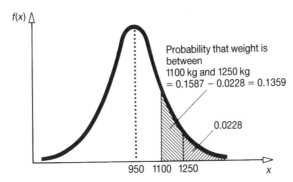

**Figure 11.12**  Calculations for part (c) of Worked Example 11.9.

For weight above 1100 kg:

$$Z = \frac{1100 - 950}{150} = 1 \quad \text{probability} = 0.1587$$

For weight above 1250 kg:

$$Z = \frac{1250 - 950}{150} = 2 \quad \text{probability} = 0.0228$$

So the probability that the weight is between these two is $0.1587 - 0.0228 = 0.1359$.

(d) To find the probability that the weight is between 800 kg and 1300 kg we use:

$P(\text{between } 800 \text{ and } 1300 \text{ kg}) = 1 - P(\text{less than } 800 \text{ kg}) - P(\text{greater than } 1300 \text{ kg})$

This is illustrated in Figure 11.13.

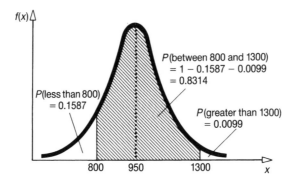

**Figure 11.13**   Calculations for part (d) of Worked Example 11.9.

For weight below 800 kg:

$$Z = \frac{800 - 950}{150} = -1 \quad \text{probability} = 0.1587$$

For weight above 1300 kg:

$$Z = \frac{1300 - 950}{150} = 2.33 \quad \text{probability} = 0.0099$$

So the probability that the weight is between these two is
$1 - 0.1587 - 0.0099 = 0.8314$.

(e)   To find the weight that 95% of cattle are below we work the other way around. If you look up 0.05 in the body of the normal tables in Appendix C you find this corresponds to 1.645 (half way between 1.64 and 1.65). This shows that 5% of cattle weigh more than 1.645 standard deviations above the mean, or:

$$P(\text{more than } X) = 0.05$$

giving

$$Z = \frac{X - 950}{150} = 1.645$$

so:

$$X - 950 = 1.645 \times 150$$

$$X = 1196.75 \text{ kg}$$

# we Worked example 11.10

At the end of section 11.4 we mentioned an example of motor insurance, where 100 drivers holding the insurance policy have an average of 0.2 accidents each a year. Use a normal distribution to find the probability that fewer than 15 drivers have accidents in a year. How could you take into account the fact that accidents are discrete?

## Solution

This is a binomial process with $n = 100$ and $p = 0.2$, so the mean $= np = 100 \times 0.2 = 20$. The standard deviation of a binomial distribution is $\sqrt{npq} = \sqrt{16} = 4$. We cannot use a Poisson approximation for the binomial, but we can use a normal approximation. Then to find the probability of fewer than 15 drivers having an accident:

$$Z = \frac{15 - 20}{4} = -1.25 \qquad \text{probability} = 0.1056$$

Because the number of accidents is discrete, a 'continuity correction' is sometimes used. We are looking for the probability of fewer than 15 accidents but it is clearly impossible for the number of accidents to be *between* 14 and 15, so we can add an allowance to interpret 'less than 15' as 'less than 14.5'. (See Figure 11.14.)

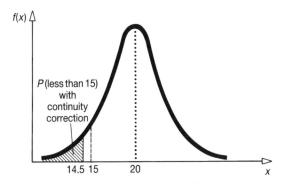

**Figure 11.14**   Normal distribution with continuity correction for Worked Example 11.10.

This continuity correction for integer values then gives

$$Z = \frac{14.5 - 20}{4} = -1.375 \qquad \text{probability} = 0.0846$$

If the question had asked for 15 or less accidents, the continuity correction could interpret this as less than 15.5. Then:

$$Z = \frac{15.5 - 20}{4} = -1.125 \qquad \text{probability} = 0.1303$$

## In summary

**Large numbers of observations often follow a normal distribution. This is a continuous distribution which gives the probability that observations are within specified ranges.**

## Chapter review

This chapter described probability distributions. In particular, it:

- discussed probability distributions in terms of relative frequencies;
- used combinations and permutations in sequencing problems;
- described the binomial distribution for trials that end in success or failure;
- described the Poisson distribution for random events;
- described the normal distribution for large numbers of observations.

## Problems

11.1 Describe the probability distribution of the following set of observations.

> 10 14 13 15 16 12 14 15 11 13 17 15 16 14 12 13 11 15 15 14
> 12 16 14 13 13 14 13 12 14 15 16 14 11 14 12 15 14 16 13 14

11.2 Find the value of $^nC_r$ and $^nP_r$ when:

(a) $r = 5$ and $n = 15$

(b) $r = 2$ and $n = 10$

(c) $r = 8$ and $n = 10$

11.3 (a) A salesman wants to visit twelve customers. In how many different orders can he visit them?

(b) One day the salesman can visit only eight customers. In how many different ways can he choose the eight?

(c) As the salesman has to travel between customers, the order in which he visits them is important. How many different schedules are there for eight customers?

11.4 An oil company drilling exploratory wells in Northern Canada, finds that 90% of the wells are dry wells and 10% are producer wells.

(a) If 12 wells are drilled, what is the probability that all 12 will be producer wells?

(b)  What is the probability that all 12 will be dry wells?

(c)  What is the probability that exactly 1 will be a producer well?

(d)  What is the probability that at least 3 will be producer wells?

11.5   During a busy period at an airport, an average of 10 planes arrive every hour. What is the probability that 15 or more planes will arrive in an hour?

11.6   A large number of observations are found to have a mean of 120 and variance of 100. What proportion of observations is:

(a)  below 100

(b)  above 130

(c)  between 100 and 130

(d)  between 130 and 140

(e)  between 115 and 135?

11.7   The number of meals served each week in a fast-food restaurant is normally distributed with a mean of 6000 and a standard deviation of 600.

(a)  What is the probability that the number of meals served in a week is less than 5000?

(b)  What is the probability that more than 7500 are served?

(c)  What is the probability that between 5500 and 6500 meals are served?

(d)  What number of meals will the restaurant serve more than, in 90% of weeks?

# CS  Case study – Machined components

The operations manager was speaking calmly to the marketing manager. 'I said it usually takes 70 days to make a batch of these components. We have to buy parts and materials, make subassemblies, set up machines, schedule operations, make sure everything is ready to start production, then actually make the components, check them and shift them to the finished goods stores, and so on. Actually making the components involves 187 distinct steps taking a total of 20 days. The whole process usually takes 70 days, but there is a lot of variability. This batch you are shouting about is going to take about 95 days because we were busy working on other jobs, and an important machine broke down, so we had to wait for parts to be flown in from Tokyo, and so on. It is your fault you heard my estimate and then assumed I was exaggerating so you promised the customer delivery in 65 days.'

The marketing manager looked worried. 'Why didn't you rush through this important job? Why is there such variation in time? Why did the break down of one machine disrupt production by so much? What am I going to say to our customer?'

The operations manager replied, 'Let me answer your questions in order. Because I was rushing through other important jobs. The variation isn't really that much; our estimates are usually within 10 days. It is a central machine which affects the capacity of the whole plant. I can only suggest you apologise and say you will listen to the operations manager more carefully in the future.'

Despite his apparent calmness, the operations manager was concerned about the variability in production times. As an experiment, he had once tried to match capacity exactly with expected throughput. Then he found that those operations near the beginning of the process were performing reasonably well, but at the end of the process the variability seemed to have been magnified and the output times seemed to be out of control. At one point he had 8 machines in a line, each of which processed a part for 10 minutes before passing it to the next machine. Although this arrangement seemed perfectly balanced, he found that stocks of work-in-progress built up dramatically. People suggested that this was because the actual processing time could vary between 5 and 15 minutes.

The operations manager really needed to see why there is variability, how much variability there should be, what the effects are of this, how could it be reduced, what benefits reduced variability would bring, and so on. This study will need funding, and a proposal must be passed by the relevant departments. How would you write this proposal?

# 12 Sampling and testing

## Contents

## Chapter outline

This chapter talks about statistical sampling and testing. After reading the chapter you should be able to:

- understand how to use sampling
- use sampling distributions
- use samples to estimate population means
- understand the approach of hypothesis testing
- test hypotheses about population means
- use chi-squared tests

# 12.1 Using samples to find population means

Chapter 3 showed how samples can get reliable data from a few observations rather than the whole population. Suppose you want to check the quality of goods arriving from a supplier. Rather than test every unit arriving, you might test a sample of one in every ten. Then you can judge the quality of all the parts from the quality of the sample.

Sampling takes some units from the population, measures a property – like quality, weight, length, size or performance – and then estimates the value of the property for the whole population. This is called **statistical inference**.

## 12.1.1 Distribution of sample means

If you take a series of samples from any population, you would expect some variation between samples. Suppose a jam factory buys apples in boxes with a nominal weight of 10 kg. If you weighed samples of boxes you would expect the mean weight to be about 10 kg, but would not be surprised by small variations. Samples taken over consecutive days might have mean weights of 10.2 kg, 9.8 kg, 10.1 kg 10.0 kg, 9.9 kg and so on. If you continued taking samples over some period, you could build a frequency distribution of the sample means. Any distribution like this that comes from samples is called a **sampling distribution**. When the distribution describes sample means, it is called a **sampling distribution of the mean**.

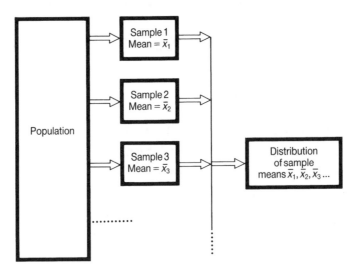

**Figure 12.1**   Derivation of the sampling distribution of the mean.

We can relate the properties of the sampling distribution of the mean back to the original population.

- If a population is normally distributed, then the sampling distribution of the mean is also normally distributed.

- If the sample size is large (more than 30), then the sampling distribution of the mean is normally distributed regardless of the population distribution.

- If the mean of the population is $\mu$, the mean of the sampling distribution of the mean is also $\mu$ (see Figure 12.2).

- If the standard deviation of the population is $\sigma$, the standard deviation of the sampling distribution of the mean is $\sigma/\sqrt{n}$, where $n$ is the sample size.

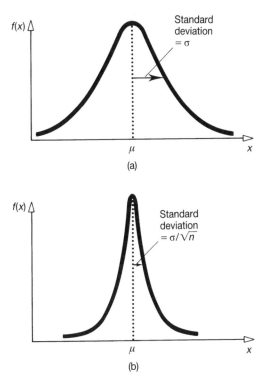

**Figure 12.2**   Comparison of distributions for (a) population and (b) sampling distribution of the mean.

An obvious problem with statistical inference is that we have to use clumsy statements to describe, for example, 'the mean of the sampling distribution of the mean'. The ideas behind these phrases are fairly simple, but you have to think clearly about what they describe.

## we Worked example 12.1

Units come off a production line with a mean length of 60 cm and standard deviation of 1 cm. What is the probability that a sample of 36 units has a mean length of less than 59.7 cm?

## Solution

If we take samples of 36 from the population, the mean length of these samples is described by the sampling distribution of the mean. This is:

● normally distributed

● has mean length $\mu = 60$ cm

● has standard deviation $\sigma/\sqrt{n} = 1/\sqrt{36} = 0.167$ cm

The probability that one sample has a mean length less than 59.7 cm comes from the area under the tail of this sampling distribution. We find this from the value of $Z$ – the number of standard deviations that the point of interest (59.7) is away from the mean:

$$Z = \frac{59.7 - 60}{0.167} = -1.80$$

Looking up 1.80 in normal tables gives the probability of 0.0359 – so we expect 3.59% of samples to have a mean length of less than 59.7 cm.

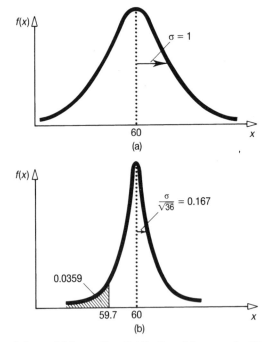

**Figure 12.3** (a) Population and (b) sampling distribution of the mean for Worked Example 12.1.

## In summary

**Statistical inference uses samples from a population. The mean of these samples is described by a sampling distribution of the mean. We can relate this distribution to the population distribution.**

### 12.1.2 Confidence intervals

In Worked Example 12.1 we found the characteristics of a sample from the known characteristics of the population. It is usually more interesting to work the other way around, and to find the characteristics of a population from a sample. Suppose we take a sample of 100 parts and find that the mean length is 30 cm. How can we estimate the mean length of the population of parts? If the sample accurately describes the population, the sample mean of 30 cm gives a reasonable estimate of the population mean. This single value is a **point estimate**.

Unfortunately, a point estimate comes from a sample, and is unlikely to be perfectly accurate. But we can define a range that the population mean is likely to be within.
We need two measures for this range:

● the limits of the range;

● a level of confidence that the mean is within the range.

Then we might say, 'we are 95% confident that the population mean is between 27 and 33 cm'. This range is called a **confidence interval**.

We can find the 95% confidence interval using the following argument. The sample mean, $\bar{x}$, is the best point estimate for the population mean, $\mu$, but this point estimate is one observation from the sampling distribution of the mean. This sampling distribution is normal, so 95% of observations are within 1.96 standard deviations of the mean. The standard deviation of the sampling distribution of the mean is $\sigma/\sqrt{n}$, so with a bit of thought you can see that:

The 95% confidence interval for the population mean is:

$$\bar{x} - 1.96\frac{\sigma}{\sqrt{n}} \quad \text{to} \quad \bar{x} + 1.96\frac{\sigma}{\sqrt{n}}$$

Similarly, the 90% confidence interval for the population mean is:

$$\bar{x} - 1.645\frac{\sigma}{\sqrt{n}} \quad \text{to} \quad \bar{x} + 1.645\frac{\sigma}{\sqrt{n}}$$

and the 99% confidence interval is:

$$\bar{x} - 2.58\frac{\sigma}{\sqrt{n}} \quad \text{to} \quad x + 2.58\frac{\sigma}{\sqrt{n}}$$

As you can see, higher levels of confidence need wider intervals.

## we Worked example 12.2

A machine makes parts with a standard deviation in length of 1.4 cm. A random sample of 100 parts has a mean length of 80 cm. What is the 95% confidence interval for the mean length of the parts?

## Solution

The point estimate for the population mean is 80 cm.

The sampling distribution of the mean has a mean of 80 cm and standard deviation of:

$$\frac{\sigma}{\sqrt{n}} = \frac{1.4}{100} = 0.14 \text{ cm}$$

Ninety-five per cent of observations are within 1.96 standard deviations of the mean, so we expect 95% of observations to be within the range:

$$\bar{x} - 1.96\frac{\sigma}{\sqrt{n}} \quad \text{to} \quad \bar{x} + 1.96\frac{\sigma}{\sqrt{n}}$$

$$80 - 1.96 \times 0.14 \quad \text{to} \quad 80 + 1.96 \times 0.14$$

that is:

$$79.73 \text{ cm} \quad \text{to} \quad 80.27 \text{ cm}$$

The results are shown in Figure 12.4.

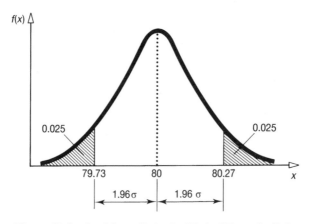

**Figure 12.4** Confidence limits for Worked Example 12.2.

In Worked Example 12.2 we estimated the population mean from a sample mean, but we assumed that we knew the standard deviation of the population. In practice, we are unlikely to know the standard deviation of a population, if we do not know its mean. So we are much more likely to use the sample to estimate both the population mean and standard deviation.

The obvious estimate of the population's standard deviation is the sample's standard deviation. Then the 95% confidence interval becomes:

$$\bar{x} - 1.96\frac{s}{\sqrt{n}} \quad \text{to} \quad \bar{x} + 1.96\frac{s}{\sqrt{n}}$$

## we Worked example 12.3

A security company employs nightwatchmen to patrol warehouses. It wants to find the average time to patrol warehouses of a certain size. On a typical night the company recorded the times to patrol 40 similar warehouses, and these had a mean of 76.4 minutes and a standard deviation of 17.2 minutes. What are the 95% and 99% confidence intervals of the population mean?

## Solution

The point estimate of the population mean is 76.4 minutes.

The standard deviation of the sample, $s$, is 17.2 minutes, and we can use this an approximation of the standard deviation of the population, $\sigma$.

The 95% confidence interval is:

$$\bar{x} - 1.96\frac{s}{\sqrt{n}} \quad \text{to} \quad \bar{x} + 1.96\frac{s}{\sqrt{n}}$$

$$76.4 - 1.96 \times \frac{17.2}{\sqrt{40}} \quad \text{to} \quad 76.4 + 1.96 \times \frac{17.2}{\sqrt{40}}$$

$$71.07 \quad \text{to} \quad 81.73$$

The 99% confidence interval is:

$$\bar{x} - 2.58\frac{s}{\sqrt{n}} \quad \text{to} \quad x + 2.58\frac{s}{\sqrt{n}}$$

$$76.4 - 2.58 \times \frac{17.2}{\sqrt{40}} \quad \text{to} \quad 76.4 + 2.58 \times \frac{17.2}{\sqrt{40}}$$

$$69.38 \quad \text{to} \quad 83.42$$

## In summary

**Interval estimates for the population mean are more useful than point estimates. We can use the sampling distribution of the mean to find confidence intervals for a population mean.**

### 12.1.3 One-sided confidence intervals

So far we have assumed that the confidence interval is symmetrical about the mean, but we are often only interested in one side of a sampling distribution. We might, for example, want to be 95% confident that the mean number of defects is below some maximum, or the weight of goods is above some minimum.

A one-sided 95% confidence interval has 5% of the distribution in one tail, as shown in Figure 12.5. This point is 1.645 standard deviations from the mean.

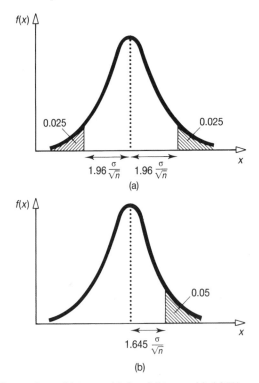

**Figure 12.5** Comparison of (a) two-sided and (b) one-sided 95% confidence interval.

Now we can use the following rules to find the one-sided 95% confidence interval:

● To find the value that we are 95% confident the population mean is above, use:

$$\bar{x} - 1.645\frac{s}{\sqrt{n}}$$

● To find the value that we are 95% confident the population mean is below, use:

$$\bar{x} + 1.645\frac{s}{\sqrt{n}}$$

We can use the same approach for other levels of confidence, and are using 95% simply because it is convenient and popular.

## we Worked example 12.4

A sample of 60 ingots from a foundry has a mean weight of 45 kg and standard deviation of 5 kg.

(a)  What weight are we 95% confident that the population mean is below?

(b)  What weight are we 95% confident the population mean is above?

## Solution

(a)  We are 95% confident that the population mean is less than $\bar{x} + 1.645 s/\sqrt{n}$, i.e.

$$45 + 1.645 \times \frac{5}{\sqrt{60}} = 46.06 \text{ kg.}$$

(b)  We are 95% confident that the population mean will be more than $\bar{x} - 1.645 s/\sqrt{n}$, i.e.

$$45 - 1.645 \times \frac{5}{\sqrt{60}} = 43.94 \text{ kg.}$$

### In summary

**Sometimes we are only interested in a confidence interval in one tail of a distribution. The overall approach is similar to the method with two-sided confidence intervals.**

# 12.2 Hypotheses about population means

## 12.2.1  Approach to hypothesis testing

In the previous section we used data from a sample to estimate values for a population. In this section we are going to test whether a belief about a population is supported by the evidence from a sample. This is the basis of **hypothesis testing**.

Suppose we have some preconceived idea about the value taken by a population variable. We might, for example, believe that domestic telephone bills have risen by 10% in the past year. This is a hypothesis we want to test. Now we can take a sample from the population and see if the results support our hypothesis or not. The formal procedure is:

- Define a simple precise statement about the situation (the hypothesis)
- Take a sample from the population
- Test this sample to see if it supports the hypothesis or if it makes the hypothesis highly improbable

# we Worked example 12.5

Bottles are filled with a nominal 400 g of fluid, but the actual weights are normally distributed with a standard deviation of 20 g. A bottle was tested and found to contain 446 g. Does the nominal weight seem reasonable?

## Solution

Following the steps listed above:

- An initial hypothesis is that the mean weight of bottles is 400 g.
- We have a very small sample that gives data to test this hypothesis.
- If the mean weight is 400 g, we can find the probability of finding a sample weighing 446 g.

The number of standard deviations from the mean, Z, is:

$$Z = \frac{446 - 400}{20} = 2.3$$

Normal tables show that this has probability = 0.01. If the hypothesis is true we are very unlikely to find a sample weighing 446 g. So we can reject the hypothesis that the mean content is 400 g, and suggest that the bottles are being overfilled.

The original statement is called the **null hypothesis**, which is usually called $H_0$. The term 'null' means that there has been no change in the value being tested since the hypothesis was formulated. In Worked Example 12.5 we rejected the null hypothesis as being very unlikely. If a null hypothesis is more likely, we can accept it, but statisticians are more cautious and usually say that they 'cannot reject the hypothesis'.

If we reject the null hypothesis then, by implication, we accept an alternative. In Worked Example 12.5 we rejected the null hypothesis that the mean weight of bottles is 400 g, so we accept the alternative hypothesis that the mean weight is not 400 g. For each null hypothesis there is always an alternative hypothesis, which is usually called $H_1$. If the null hypothesis, $H_0$, is that domestic telephone bills have risen by 10% in the last year, the alternative hypothesis, $H_1$, is that they have not risen by 10%. If the null hypothesis, $H_0$, is that first class letters take 2 days to reach their destination, the alternative hypothesis, $H_1$, is that they do not take 2 days to reach their destination.

Notice that the null hypothesis is a simple, specific statement, while the alternative hypothesis is more vague. In practice, the null hypothesis is phrased in terms of one thing equalling another. We might have a null hypothesis that the mean weight is 1.5 kg and an alternative hypothesis that the mean weight is not 1.5 kg. A null hypothesis might be that the average salary in an office is £20 000, while the alternative hypothesis is that the average salary is lower than this.

## In summary

**A null hypothesis is a precise statement about a situation. Hypothesis-testing uses a sample to see if the evidence supports this statement or if we should reject the hypothesis.**

## 12.2.2  Significance levels

Even good samples are not completely accurate. So when we use a sample to test a hypothesis, there is some uncertainty in the result. In Worked Example 12.5 we said that the result was unlikely and rejected the null hypothesis. However in 1% of samples the result we found would happen by chance, and we would reject a perfectly true hypothesis. In general, there are two ways of getting the wrong answer with hypothesis testing:

● Type I error, when we reject a null hypothesis that is true.

● Type II error, when we do not reject a null hypothesis that is false.

The different combinations are set out in Table 12.1

**Table 12.1**

|  |  | Null hypothesis is | |
| --- | --- | --- | --- |
|  |  | *True* | *False* |
| *Decision* | *Not reject* | Correct decision | Type II error |
|  | *Reject* | Type I error | Correct decision |

Ideally we would like the probability of both Type I and Type II errors to be very small. The only way of doing this is to use a large sample. If we try any other adjustments to reduce the probability of Type I errors, we increase the probability of Type II errors, and vice versa. With a limited sample size, we have to accept a compromise between the two types of error.

The way to get this compromise is to use a **significance level**, which is the minimum acceptable probability that an observation is a random sample from the hypothesised population. With a 5% significance level, we will not reject a null hypothesis if there is a probability greater than 5% that an observation comes from a population with the specified value. On the other hand, if there is a probability of less than 5% that the observation came from this population, we reject the null hypothesis.

If the sampling distribution is normal, a 5% significance level will not reject a null hypothesis if the sample result is within 1.96 standard deviations of the mean, and reject it if it is outside this range (as shown in Figure 12.6). Remember, though, that 5% of observations fall outside this range even if the null hypothesis is true. In other words, there is a 5% chance of rejecting a true null hypothesis – or making a Type I error.

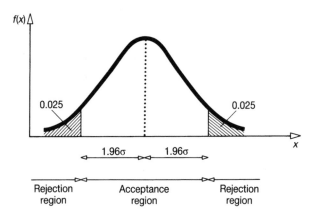

**Figure 12.6**   Acceptance and rejection regions of 5% significance level.

With a 1% significance level we do not reject the null hypothesis if an observation is within 2.58 standard deviations of the mean. So lower significance levels need stronger evidence to reject the null hypothesis. With lower significance levels the probability of a Type I error is reduced, but the probability of a Type II error is increased.

## we Worked example 12.6

Auditors suggest that the mean value of accounts received by a firm is £260. They take a sample of 36 accounts, and find that the mean is £240 with a standard deviation of £45. Use a 5% significance level to test the auditors' suggestion.

## Solution

The null hypothesis is that the mean value of accounts is £260, and the alternative hypothesis is that the mean is not £260. So:

$$H_0: \quad \mu = 260$$

$$H_1: \quad \mu \neq 260$$

The significance level is 5%, so we reject the null hypothesis if there is a probability of less 0.05 of getting this result.

     With a sample of 36, the sampling distribution of the mean is normal, with mean 260 and standard deviation $45/\sqrt{36}$. Then the acceptance range is all points within 1.96 standard deviations of the mean:

$$260 - 1.96 \times \frac{45}{\sqrt{36}} \quad \text{to} \quad 260 + 1.96 \times \frac{45}{\sqrt{36}}$$

which is

$$245.3 \quad \text{to} \quad 274.7$$

The actual observation is outside this range, so we reject the null hypothesis and accept the alternative hypothesis that the mean value of accounts is not equal to £260.

Now we have the detailed steps in hypothesis testing.

- State the null and alternative hypothesis
- Specify the level of significance to be used
- Calculate the acceptance range for the variable tested
- Find the actual value for the variable tested
- See whether or not to reject the null hypothesis
- State the conclusion reached

## we Worked example 12.7

The average price of a car in a certain city is claimed to be £15 000. A sample of 45 cars had a mean price of £14 300 with a standard deviation of £2000.

(a) Use a 5% significance level to check the claim.

(b) What is the effect of a 1% significance level?

## Solution

(a) We follow the procedure described above.

- State the null and alternative hypotheses.

  $$H_0: \quad \mu = 15\,000 \qquad H_1: \quad \mu \neq 15\,000$$

- Specify the level of significance to be used.

  This is 5%.

- Calculate the acceptance range for the variable tested.

  With a sample of 45, the sampling distribution of the mean is normal with mean 15 000 and standard deviation $2000/\sqrt{45} = 298.14$. For a 5% significance level we look at the points that are within 1.96 standard deviations of the mean (see Figure 12.7). So the acceptance range is:

  $$15\,000 - 1.96 \times 298.14 \quad \text{to} \quad 15\,000 + 1.96 \times 298.14$$

  or $\qquad\qquad\qquad\qquad 14\,416 \quad \text{to} \quad 15\,584$

- Find the actual value for the variable tested

  This is £14 300.

- See whether or not to reject the null hypothesis.

  The actual value is outside the acceptance range, so we must reject the null hypothesis.

- State the conclusion reached.

  The evidence from the sample does not support the original claim that the average price of a car in the city is £15 000.

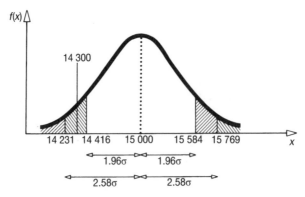

**Figure 12.7**  Acceptance ranges for Worked Example 12.7.

(b) With a 1% significance level, the acceptance range is within 2.58 standard deviations of the mean, or:

$$15\,000 - 2.58 \times 298.14 \quad \text{to} \quad 15\,000 + 2.58 \times 298.14$$

which is:

$$14\,231 \quad \text{to} \quad 15\,769$$

The actual observation is £14\,300 which is within this range, and we cannot reject the null hypothesis.

## In summary

**A significance level is the minimum acceptable probability that an observation is a random sample from the hypothesised population. It is equivalent to the probability of making a Type I error.**

## 12.2.3 One-sided tests

The problems we have looked at so far have had a null hypothesis of the form:

$$H_0: \quad \mu = 10$$

and an alternative hypothesis in the form:

$$H_1: \quad \mu \neq 10$$

In practice, we are often concerned that a value is above (or sometimes below) the claimed value. If we buy boxes of chocolates, we only want to make sure that their weight is not below the specified value; on the other hand, if we are delivering parcels, we only want to make sure their weight is not above the claimed value. We can tackle this kind of problem by adjusting the alternative hypothesis.

If you are buying boxes of chocolates with a claimed weight of 500 g, and want to make sure that the actual weight is not below this, you can have:

Null hypothesis, $H_0$:  $\mu = 500$ g

Alternative hypothesis, $H_1$:  $\mu < 500$ g

If you are delivering parcels with a claimed weight of 25 kg, and want to make sure the actual weight is not above this, you can have:

Null hypothesis, $H_0$:  $\mu = 25$ kg

Alternative hypothesis, $H_1$:  $\mu > 25$ kg

For this kind of test we only use one tail of the sampling distribution. So a 5% significance level has the 5% area of rejection in one tail of the distribution. In a normal distribution this point is 1.645 standard deviations from the mean, as shown in Figure 12.8.

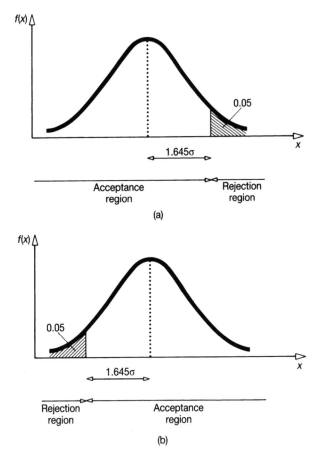

**Figure 12.8** One-sided tests for 5% significance level:
(a) when concerned with a maximum value; (b) when
concerned with a minimum value.

## we Worked example 12.8

A mail order company charges a flat rate for delivery based on a mean weight for packages of 1.75 kg with a standard deviation of 0.5 kg. Postal charges now seem high, and a random sample of 100 packages has a mean weight of 1.86 kg. What does this show?

## Solution

We use the standard procedure, as follows.

● State the null and alternative hypotheses.

The company want to see if the mean weight of packages is above 1.75 Kg, so:

$$H_0: \quad \mu = 1.75 \text{ kg} \qquad H_1: \quad \mu > 1.75 \text{ kg}$$

● Specify the level of significance to be used.

We will assume this is 5%.

● Calculate the acceptance range for the variable tested.

With a sample of 100, the sampling distribution of the mean is normal with mean of 1.75 kg, standard deviation $0.5/\sqrt{100} = 0.05$ kg. For a 5% significance level and a one-sided test, we look at the points that are more than 1.645 standard deviations above the mean. The acceptance range is below:

$$1.75 + 1.645 \times 0.05 = 1.83 \text{ kg}$$

● Find the actual value for the variable tested.

The mean weight of the sample of parcels is 1.86 kg.

● See whether or not to reject the null hypothesis.

The actual value is outside the acceptance range, so we must reject the null hypothesis.

● State the conclusion reached.

The sample does not support the view that the mean weight of packages is 1.75 kg, but suggests that it is more than this (see Figure 12.9).

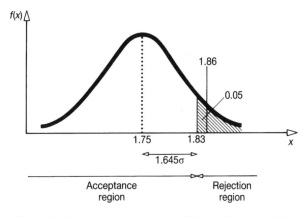

**Figure 12.9**   Acceptance range for Worked Example 12.8.

## In summary

**We can extend the standard two-sided hypothesis test to a one-sided test when we want to see whether means are above or below specified values.**

# 12.3 Chi-squared test

We have often said something like, 'the data follow a normal distribution' – But how do we know this? Here we are going to describe a test to see if data follow an expected distribution. This test is the $\chi^2$ test – where $\chi$ is the Greek letter chi.

The chi-squared test looks at the frequency of observations and sees if these match expected frequencies.

Suppose we have a series of observed frequencies $O_1$, $O_2$, $O_3$, ..., $O_n$, and are expecting the frequencies $E_1$, $E_2$, $E_3$, ..., $E_n$. The difference between these, $(O_i - E_i)$, tells us how closely the observations match expectations. Squaring the difference removes any negative values, and then dividing by $E_i$ gives a distribution with a standard shape. So we will calculate:

$$\chi^2 = \frac{(O_1 - E_1)^2}{E_1} + \frac{(O_2 - E_2)^2}{E_2} + \frac{(O_3 - E_3)^3}{E_3} + \dots + \frac{(O_n - E_n)^2}{E_n}$$

or

$$\chi^2 = \sum \frac{(O - E)^2}{E}$$

This shape of this distribution depends on the degrees of freedom – which we will describe later – as shown in Figure 12.10.

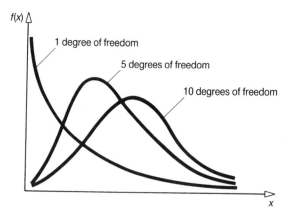

**Figure 12.10**   Chi-squared distribution with varying degrees of freedom.

If the observed frequencies are close to the expected frequencies, $\chi^2$ will be close to zero. But if the differences are large $\chi^2$ will be bigger. So we will find a test value of $\chi^2$ – called the **critical value** – that is the biggest value we will accept. If $\chi^2$ is above this critical value we say that actual values do not match expected ones; but if $\chi^2$ is below the critical value then the observations match the expected values.

You can look up critical values for $\chi^2$ in the tables in Appendix D. These depend on the degrees of freedom, which are defined as:

degrees of freedom = number of classes – number of estimated variables – 1

This seems a rather strange idea, but we need it to allow for the number of classes we are using for the data. Then we can use the $\chi^2$ values in the usual procedure for hypothesis testing.

## we Worked example 12.9

Five factories report the following numbers of accidents in a five-year period. Do these figures suggest that some factories have more accidents than others?

| Factory | 1 | 2 | 3 | 4 | 5 |
|---|---|---|---|---|---|
| Number of accidents | 31 | 42 | 29 | 35 | 38 |

## Solution

We can use the standard procedure.

● State the null and alternative hypotheses.

The null hypothesis, $H_o$, is that each factory has the same number of accidents. The alternative hypothesis, $H_1$, is that each factory does not have the same number of accidents.

● Specify the level of significance to be used.

We will take this as 5%.

● Calculate the critical value of $\chi^2$.

In this problem there are five classes, and no variables have been estimated, so the degrees of freedom are $5 - 0 - 1 = 4$. With a 5% significance level we look up 0.05 in $\chi^2$ tables and find a critical value of 9.49.

● Find the actual value of $\chi^2$.

For this we calculate:

$$\chi^2 = \sum \frac{(O - E)^2}{E}$$

There are 175 accidents. If each factory has the same number, there would be $E = 175/5 = 35$ accidents in each. Then Table 12.2 shows the calculations that we need.

**Table 12.2**

| Factory | O | E | (O − E) | (O − E)² | (O − E)²/E |
|---------|-----|-----|------|------|------|
| 1 | 31 | 35 | −4 | 16 | 0.46 |
| 2 | 42 | 35 | 7 | 49 | 1.40 |
| 3 | 29 | 35 | −6 | 36 | 1.03 |
| 4 | 35 | 35 | 0 | 0 | 0.00 |
| 5 | 38 | 35 | 3 | 9 | 0.26 |
| Totals | 175 | 175 | | | 3.15 |

The actual value of $\chi^2$ is 3.15.

● See whether or not to reject the null hypothesis.

The actual value of 3.15 is less than the critical value of 9.49 so we cannot reject the null hypothesis.

● State the conclusion reached.

We cannot say that factories have different numbers of accidents. The variations are probably due to chance.

## we Worked example 12.10

A company records the following number of defective reports from an office each day. Do these numbers follow a Poisson distribution?

| Defects | 0 | 1 | 2 | 3 | 4 | 5 |
|---------|----|----|----|----|----|----|
| Days | 21 | 43 | 19 | 11 | 4 | 2 |

## Solution

Again we use the standard procedure.

● State the null and alternative hypotheses.

The null hypothesis, $H_o$, is that the distribution is Poisson. The alternative hypothesis, $H_1$, is that the distribution is not Poisson.

● Specify the level of significance to be used.

We can take this as 5%.

- Calculate the critical value of $\chi^2$.

    In this problem there is going to be some adjustment of the data, so we will return to this calculation a little later.

- Find the actual value of $\chi^2$.

    We have 100 days of data, and the total number of defects is:

    $$(21 \times 0) + (43 \times 1) + (19 \times 2) + (11 \times 3) + (4 \times 4) + (2 \times 5) = 140$$

    The mean number of defects is $140/100 = 1.4$ a day. We can use Appendix B to give the probability of 0, 1, 2, 3, 4 and 5 defects each day with a Poisson distribution. Multiplying these probabilities by the number of days gives the expected distribution of defects shown in the following table.

| Number of defects | 0 | 1 | 2 | 3 | 4 | 5 |
|---|---|---|---|---|---|---|
| Probability | 0.2466 | 0.3452 | 0.2417 | 0.1128 | 0.0395 | 0.0111 |
| Expected frequency | 24.66 | 34.52 | 24.17 | 11.28 | 3.95 | 1.11 |

One problem here is that the $\chi^2$ distribution does not work well with expected frequencies of less than 5. So we have to combine adjacent classes to give expected number of observations greater than 5. Redefining the last class as '4 or more' gives the following expected distribution.

| Number of defects | 0 | 1 | 2 | 3 | 4 or more |
|---|---|---|---|---|---|
| Probability | 0.2466 | 0.3452 | 0.2417 | 0.1128 | 0.0537 |
| Expected frequency | 24.66 | 34.52 | 24.17 | 11.28 | 5.37 |

Now you can see why we delayed the calculation of the critical value. There are now five classes, and we have estimated one parameter – the mean number of successes – so the number of degrees of freedom is:

number of classes − number of estimated parameters − $1 = 5 - 1 - 1 = 3$

Looking up the critical value for a significance level of 5% (0.05) and 3 degrees of freedom gives a value of 7.81.

The actual value of $\chi^2$ is 3.81 as calculated in Table 12.3.

**Table 12.3**

| Frequency | O | E | (O − E) | (O − E)$^2$ | (O − E)$^2$/E |
|---|---|---|---|---|---|
| 0 | 21 | 24.66 | −3.66 | 13.40 | 0.54 |
| 1 | 43 | 34.52 | 8.48 | 71.91 | 2.08 |
| 2 | 19 | 24.17 | −5.17 | 26.73 | 1.11 |
| 3 | 11 | 11.28 | −0.28 | 0.08 | 0.01 |
| 4 or more | 6 | 5.37 | 0.63 | 0.40 | 0.07 |
| **Totals** | **100** | **100** | | | **3.81** |

- See whether or not to reject the null hypothesis.

  The actual value of $\chi^2$ is 3.81 which is less than the critical value of 7.81 so we cannot reject the null hypothesis.

- State the conclusion reached.

  The evidence suggests that the observations follow a Poisson distribution.

## we Worked example 12.11

A factory is doing some statistical analyses on the mean weight of a product, but these analyses are only valid if the weights are normally distributed. The product has a mean weight of 45 g and a standard deviation of 15 g. A sample of 500 units was taken, with weight distribution given in Table 12.4. Does this suggest that the weights are normally distributed?

**Table 12.4**

| Weight (grammes) | Number of observations |
|---|---|
| less than 10 | 9 |
| 10 – 19.99 | 31 |
| 20 – 29.99 | 65 |
| 30 – 39.99 | 97 |
| 40 – 49.99 | 115 |
| 50 – 59.99 | 94 |
| 60 – 69.99 | 49 |
| 70 – 79.99 | 24 |
| 80 – 89.99 | 16 |

## Solution

This is an important question as many statistical analyses are only valid if there is a normal distribution.

- State the null and alternative hypotheses.

  The null hypothesis, $H_0$, is that the distribution is normal. The alternative hypothesis, $H_1$, is that the distribution is not normal.

- Specify the level of significance to be used.

  We can take this as 5%.

- Calculate the critical value of $\chi^2$.

    The number of degrees of freedom is $(9 - 0 - 1 =) \ 8$. Then Appendix D shows that the critical value for $\chi^2$ with 8 degrees of freedom at a 5% significance level is 15.51.

- Find the actual value of $\chi^2$.

    The probability an observation is in the range 10 to 19.99 is:

    $$P(\text{between 10 and 19.99}) = P(\text{less than 20}) - P(\text{less than 10})$$

    Now, 20 is $(20 - 45)/15 = -1.67$ standard deviations from the mean, corresponding to a probability of 0.0475, and 10 is $(10 - 45)/15 = -2.33$ standard deviations from the mean, corresponding to a probability of 0.0099. Hence:

    $$P(\text{between 10 and 19.99}) = 0.0475 - 0.0099 = 0.0376$$

    The expected number of observations in this range is $0.0376 \times 500 = 18.8$. Repeating this calculation for the other probabilities gives the results set out in Table 12.5.

    **Table 12.5**

    | Weight | O | Probability | E | $(O - E)^2/E$ |
    |---|---|---|---|---|
    | < 10 | 9 | 0.0099 | 4.95 | 3.31 |
    | 10 – 19.99 | 31 | 0.0376 | 18.8 | 7.92 |
    | 20 – 29.99 | 65 | 0.1112 | 55.6 | 1.59 |
    | 30 – 39.99 | 97 | 0.2120 | 106.0 | 0.76 |
    | 40 – 49.99 | 115 | 0.2586 | 129.3 | 1.58 |
    | 50 – 59.99 | 94 | 0.2120 | 106.0 | 1.36 |
    | 60 – 69.99 | 49 | 0.1112 | 55.6 | 0.78 |
    | 70 – 79.99 | 24 | 0.0376 | 18.8 | 1.44 |
    | 80 – 89.99 | 16 | 0.0099 | 4.95 | 24.67 |
    | **Totals** | **500** | **1.0000** | **500** | **43.41** |

    The value of $\chi^2$ is 43.41.

- See whether or not to reject the null hypothesis.

    The actual value of 43.41 is greater than the critical value of 15.5, so we must reject the hypothesis that the sample is normally distributed.

- State the conclusion reached.

    The evidence suggests that the observations do not follow a normal distribution.

## In summary

We can use the $\chi^2$ distribution in hypothesis testing. It is most widely used to see if data follow a specified distribution.

# Chapter Review

This chapter has described two important types of statistical analysis: sampling and hypothesis testing. In particular, it:

- outlined the purpose of sampling and statistical inference;
- described the sampling distribution of the mean;
- found point estimates and confidence intervals for population means;
- used one-sided confidence tests;
- discussed the approach of hypothesis testing;
- used hypothesis tests for population means;
- extended the approach to one-sided tests;
- used chi-squared tests for goodness of fit.

# Problems

12.1 A production line makes units with a mean weight of 80 g and standard deviation of 5 g. What is the probability that a sample of 100 units has a mean weight of less than 79 g?

12.2 A machine produces parts with a variance of 14.5 cm in length. A random sample of 50 parts has a mean length of 106.5 cm. What are the 95% and 99% confidence intervals for the length of all parts?

12.3 During an audit, a random sample of 60 invoices was taken from a large population. The mean value of invoices in this sample was £125.50 and the standard deviation was £10.20. Find the 90% and 95% confidence intervals for the mean value of all invoices.

12.4 A customer feels that the amount of cereal in a particular box seems to have decreased. To test this, the customer takes a sample of 40 boxes and finds that the mean weight is 228 g with a standard deviation of 11 g.

(a) What weight is the customer 95% confident that the mean falls below?

(b) What are the two-sided confidence limits on this weight?

12.5 The mean wage of people living in a block of flats is said to be £400 a week with a standard deviation of £100. A random sample of 36 people was examined.

(a) What is the acceptance range for a 5% significance level?

(b) What is the acceptance range for a 1% significance level?

12.6 The weight of packets of biscuits is claimed to be 500 g. A random sample of 50 packets has a mean weight of 495 g and a standard deviation of 10 g. Use a significance level of 5% to see if the data from the sample support the original claim.

12.7 A television has an advertised life of 30 000 hours. A sample of 50 sets had a life of 28 500 hours with a standard deviation of 1 000 hours. What can be said about the advertisements?

12.8 Five factories report the following numbers of accidents in a year.

| Factory | 1 | 2 | 3 | 4 | 5 |
|---|---|---|---|---|---|
| Number of Accidents | 23 | 45 | 18 | 34 | 28 |

Does this evidence suggest that some factories have more accidents than others?

12.9 The number of road accidents reporting to a hospital emergency ward is shown in the following table. Do these figures follow a Poisson distribution?

| Number of Accidents | 0 | 1 | 2 | 3 | 4 | 5 | 6 |
|---|---|---|---|---|---|---|---|
| Number of days | 17 | 43 | 52 | 37 | 20 | 8 | 4 |

12.10 Do the figures in Table 12.6 follow a normal distribution?

**Table 12.6**

| Weight (grammes) | Number of observations |
|---|---|
| less than 5 | 5 |
| 5 – 19.99 | 43 |
| 20 – 34.99 | 74 |
| 35 – 49.99 | 103 |
| 50 – 64.99 | 121 |
| 65 – 79.99 | 97 |
| 80 – 94.99 | 43 |
| 95 – 109.99 | 21 |
| 110 and more | 8 |

# CS Case study – King's Fruit Farm

In 1987 James Filbert became the latest manager of King's Farm in Cambridgeshire. This farm is over 3000 acres in extent and is owned by a private agricultural company. King's Fruit Farm is a subsidiary of King's Farm and owns a number of apple, pear, plum and cherry orchards.

Recently, James has been looking at the sales of plums. These are graded and sold as fresh fruit to local shops and markets, for canning to a local cannery, or for jam to a more distant processor. The plums sold for canning generate about half as much income as those sold for fruit, but twice as much income as those sold for jam.

James is trying to estimate the weight of plums sold each year. He does not know this, as the plums are sold by the basket rather than by weight. Each basket holds about 25 kg of plums. For a pilot study, James set up some scales to see if he could weigh the amount of fruit in a sample of baskets. On the first day he weighed 10 baskets, 6 of which were sold as fresh fruit, 3 for canning and 1 for jam. The weights of fruit, in kilogrammes, were as follows:

25.6   20.8   29.4   28.0   22.2   23.1   25.3   26.5   20.7   21.9

This trial seemed to work, so James then weighed a sample of 50 baskets on three consecutive days. The weights of fruit, in kilogrammes, were as follows:

| Day 1 | 24.6 | 23.8 | 25.1 | 26.7 | 22.9 | 23.6 | 26.6 | 25.0 | 24.6 | 25.2 |
|-------|------|------|------|------|------|------|------|------|------|------|
|       | 25.7 | 28.1 | 23.0 | 25.9 | 24.2 | 21.7 | 24.9 | 27.7 | 24.0 | 25.6 |
|       | 26.1 | 26.0 | 22.9 | 21.6 | 28.2 | 20.5 | 25.8 | 22.6 | 30.3 | 28.0 |
|       | 23.6 | 25.7 | 27.1 | 26.9 | 24.5 | 23.9 | 27.0 | 26.8 | 24.3 | 19.5 |
|       | 31.2 | 22.6 | 29.4 | 25.3 | 26.7 | 25.8 | 23.5 | 20.5 | 18.6 | 21.5 |
| Day 2 | 26.5 | 27.4 | 23.8 | 24.8 | 30.2 | 28.9 | 23.6 | 27.5 | 19.5 | 23.6 |
|       | 25.0 | 24.3 | 25.3 | 23.3 | 24.0 | 25.1 | 22.2 | 20.1 | 23.6 | 25.8 |
|       | 24.9 | 23.7 | 25.0 | 24.9 | 27.2 | 28.3 | 29.1 | 22.1 | 25.0 | 23.8 |
|       | 18.8 | 19.9 | 27.3 | 25.6 | 26.4 | 28.4 | 20.8 | 24.9 | 25.4 | 25.6 |
|       | 24.9 | 25.0 | 24.1 | 25.5 | 25.2 | 26.8 | 27.7 | 20.6 | 31.3 | 29.5 |
| Day 3 | 27.2 | 21.9 | 30.1 | 26.9 | 23.5 | 20.7 | 26.4 | 25.1 | 25.7 | 26.3 |
|       | 18.0 | 21.0 | 21.9 | 25.7 | 28.0 | 26.3 | 25.9 | 24.7 | 24.9 | 24.3 |
|       | 23.9 | 23.0 | 24.1 | 23.6 | 21.0 | 24.6 | 25.7 | 24.7 | 23.3 | 22.7 |
|       | 22.9 | 24.8 | 22.5 | 26.8 | 27.4 | 28.3 | 31.0 | 29.4 | 25.5 | 23.9 |
|       | 29.5 | 23.3 | 18.6 | 20.6 | 25.0 | 25.3 | 26.0 | 22.2 | 23.9 | 25.7 |

He also recorded the number of each sample sent to each destination (Table 12.7).

**Table 12.7**

|       | Fruit | Cans | Jam |
|-------|-------|------|-----|
| Day 1 | 29    | 14   | 7   |
| Day 2 | 25    | 15   | 10  |
| Day 3 | 19    | 15   | 16  |

Pickers are paid by the basket, so the payments book was used to find the number of baskets picked on the three days as 820, 750 and 700 respectively. During a good harvest, around 6000 baskets are picked.

What information can James find from these figures?

# 13 Analysing business decisions

## Contents

## Chapter outline

This chapter shows how to tackle decisions in a variety of circumstances. After reading the chapter you should be able to:

- draw a map of a decision situation
- build a payoff matrix
- make decisions under certainty
- describe situations of uncertainty and use decision criteria
- describe situations of risk and use expected values
- use Bayes' theorem to update probabilities for decisions under risk
- discuss the use of utilities
- use decision trees for sequential decisions

# 13.1 Giving structure to decisions

Everybody has to make decisions: which car is the best buy; where should we eat; should we drive to work or go by train; should we buy a house, and so on. Most of these decisions are not very important, and we can use a combination of experience and intuition. Business decisions, however, can be very important, and managers need a more formal approach to decision making.

Suppose a company makes too little profit on one of its products. Two obvious remedies are either to reduce costs or to increase the price. If the company increases the price, though, the demand may fall. On the other hand, reducing the costs might reduce the price to customers and increase demand. If demand changes, the factory may have to reschedule production and change marketing strategies. Changed production schedules can affect production of other items, change employment prospects and so on.

We could continue with these more or less random thoughts for some time but would soon lose track of our arguments. It would be useful to have a simple diagram to summarise our thinking. We can use a **problem map** for this, and Figure 13.1 shows part of a map for the discussion above.

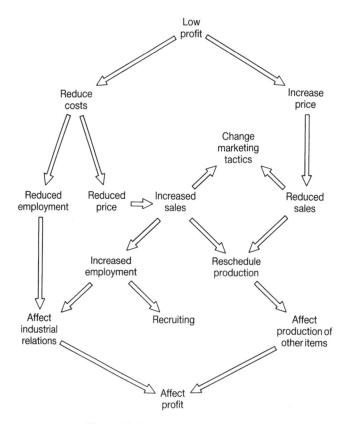

**Figure 13.1**   Part of a problem map.

This map gives an informal way of presenting connected ideas. A more formal approach to decisions looks at the features that are common to all decisions. These are:

- A decision maker, who is responsible for making the decision
- A number of alternatives, one of which must be chosen
- An objective of choosing the best alternative
- After the decision has been made, events occur over which the decision maker has no control
- Each combination of an alterntive chosen followed by events happening, leads to an outcome that has a measurable value

Consider the much-simplified example of a house owner who is offered fire insurance at a cost of £400 a year. The decision maker is the person who owns the house has an objective of minimising costs and must choose the best alternative from:

- insure the house, or
- do not insure the house.

Then an event happens, but the decision maker has no control over whether it is:

- the house burns down, or
- the house does not burn down.

The outcomes are the costs of each combination of decision and event. If the value of the house is £100 000 we can summarise this problem in Table 13.1.

**Table 13.1**

|  |  | Event | |
|---|---|---|---|
|  |  | *House burns down* | *House does not burn down* |
| *Alternatives* | *Insure house* | £400 | £400 |
|  | *Do not insure house* | £100 000 | £0 |

Table 13.1 is a **payoff matrix** and the entries show the cost to the house owner of every combination of alternative and event. Remember that the rows show alternatives – one of which is chosen – and the columns show events – one of which happens, but the decision maker does not know in advance which one. The outcomes given in the body of the matrix can be either costs or gains.

## In summary

**Managers make decisions in complex situations. Maps and payoff matrices show the structure of problems.**

# 13.2 Decisions with certainty

With decision making under certainty, we know – with certainty – which event will happen. As there is only one event to consider, we can list all the outcomes and choose the alternative which is best.

Suppose you have £1000 to invest for a year. All the alternatives you want to consider, and the outcomes, may be given in the payoff matrix shown in Table 13.2.

**Table 13.2**

|  |  | Event |
|---|---|---|
|  |  | Earns interest to give at the end of the year |
|  | Bank | £1065 |
|  | Building society | £1075 |
| Alternatives | Government stock | £1085 |
|  | Stock market | £1100 |
|  | Others | £1060 |

There is only one event – earns interest – and by looking down the list of outcomes you can identify the best. The highest value at the end of the year comes from investing in the stock market.

## we Worked example 13.1

A restaurant manager has a booking for a wedding banquet. He has a number of ways of getting staff – paying full-time staff to work overtime (costing £400), hiring current part-time staff for the day (£300), hiring new temporary staff (£350) or using an agency (£550). Draw a payoff matrix for this decision and find the best alternative.

## Solution

The payoff matrix for this decision under certainty is shown in Table 13.3. The entries are costs, so we want the lowest. This is £300 for hiring current part-time staff for the day.

**Table 13.3**

|  |  | Event |
|---|---|---|
|  |  | Pay staff |
|  | Pay full-time staff for overtime | 400 |
| Alternatives | Hire current part-time staff | 300 |
|  | Hire new temporary staff | 350 |
|  | Use an agency | 500 |

In reality, even decisions under certainty can be difficult. Would it be better, for example, for the National Health Service to invest money in buying more kidney dialysis machines, giving nurses higher wages, doing open heart surgery, funding research into cancer, or providing more parking spaces at hospitals?

## In summary

**Decisions with certainty have only one event. To choose the best alternative we find the best outcome. In practice, this may be quite difficult.**

# 13.3 Decisions with uncertainty

Most decisions have a number of events that may happen. Often we cannot say which event will actually occur, or even give reasonable probabilities. If someone changes her job a number of events may happen: she may not like the new job and quickly start looking for another, she may get the sack, she may like the job and stay, she might be moved by the company. These events are outside the control of the person taking the new job and it is impossible to give reliable probabilities.

When we cannot give the events probabilities, we are dealing with **uncertainty**. Then simple rules called **decision criteria** are useful for recommending a decision. There are many different criteria, and we will illustrate three common ones.

## 13.3.1  Laplace decision criterion

As we cannot give probabilities to events, Laplace suggests that we treat them as equally likely, and find the best average outcome. The procedure is:

- Find the average outcome for each alternative (i.e. the average of each row in the payoff matrix).

- Choose the alternative with the best average outcome (i.e. lowest cost or highest gain).

## **we** Worked example 13.2

A restaurateur is going to set up a cream tea stall at a local gala. On the morning of the gala she visits the wholesale market and has to decide whether to buy a large, medium or small quantity of strawberries. Her profit depends on the number of people attending the gala, and this is set by the weather. The matrix of gains (in thousands of pounds) for different weather conditions is given in Table 13.4. What quantity of strawberries should the restaurateur buy?

**Table 13.4**

|  |  | Event |  |  |
|---|---|---|---|---|
|  |  | Weather good | Weather average | Weather poor |
| Alternatives | Large quantity | 10 | 4 | −2 |
|  | Medium quantity | 7 | 6 | 2 |
|  | Small quantity | 4 | 1 | 4 |

## Solution

Take the average outcome for each alternative:

Large quantity $(10 + 4 - 2)/3 = 4$

Medium quantity $(7 + 6 + 2)/3 = 5$

Small quantity $(4 + 1 + 4)/3 = 3$

Choose the best average outcome. As these figures are profits, the best is the highest, which is to buy a medium quantity.

### In summary

**Decision criteria are simple rules for helping with decisions under uncertainty. The Laplace criterion finds the average outcome for each alternative and then chooses the alternative with the best average outcome.**

### 13.3.2 Wald decision criterion

Most organisations have limited resources and cannot risk a large loss. The Wald decision criterion avoids large potential losses, and suggests decisions that are cautious or even pessimistic. The steps are:

- For each alternative, find the worst outcome.

- Choose the alternative with the best of these worst outcomes.

With a payoff matrix showing costs, this is sometimes known as the **minimax cost** criterion as it looks for the maximum cost of each alternative and then chooses the alternative with the minimum of these – so it looks for the minimum maximum cost or minimax cost.

## we Worked example 13.3

Use the Wald decision criterion for the example of the cream tea stall described in Worked Example 13.2.

|  | | Event | | |
|---|---|---|---|---|
|  | | Weather good | Weather average | Weather poor |
| Alternatives | Large quantity | 10 | 4 | −2 |
|  | Medium quantity | 7 | 6 | 2 |
|  | Small quantity | 4 | 1 | 4 |

## Solution

Take the worst outcome for each alternative:

Large quantity    Minimum of $[10, 4, -2] = -2$

Medium quantity    Minimum of $[7, 6, 2] = 2$

Small quantity    Minimum of $[4, 1, 4] = 1$

Choose the best of these worst outcomes. As the figures are profits, the best is the highest, which comes from buying a medium quantity.

### In summary

**The Wald decision criterion is pessimistic and assumes that the worst outcome will happen. Then it chooses the alternative with the best of these worst outcomes.**

### 13.3.3  Savage decision criterion

Sometimes we are judged not by how well we actually did, but by how well we could have done. A student who gets 70% in an exam might be judged by the fact that he did not get 100%. An investment broker who recommends an investment in platinum may be judged not by the fact that platinum rose 15% in value, but by the fact that gold rose 25%.

In these cases there is a **regret** which is the difference between actual outcome and best possible outcome. A student who gets 70% has a regret of $100 - 70 = 30\%$. Investors who gain 15% when they could have gained 25% have a regret of $25 - 15 = 10\%$. The Savage criterion uses these regrets in the following steps.

- For each *event*, find the best possible outcome (i.e. find the best entry in each column of the payoff matrix).

- Find the regret for every other entry in the column, which is the difference between the best in the column and the other entry.

- Put the regrets into a 'regret matrix'. There is at least one zero in each column and regrets are always positive.

- For each alternative find the highest regret (i.e. the largest number in each row).

- Choose the alternative with the lowest of these highest regrets.

As you can see these last two steps apply the Wald criterion to the regret matrix.

# we Worked example 13.4

Use the Savage criterion on the example of the cream tea stall of Worked Example 13.2.

## Solution

Using the procedure described above.

- The best outcome for each event is underlined in Table 13.5 (i.e. with good weather a large quantity, with average weather a medium quantity and with poor weather a small quantity).

**Table 13.5**

|  |  | Event | | |
|---|---|---|---|---|
|  |  | Weather good | Weather average | Weather poor |
| Alternatives | Large quantity | <u>10</u> | 4 | -2 |
|  | Medium quantity | 7 | <u>6</u> | 2 |
|  | Small quantity | 4 | 1 | <u>4</u> |

- The regret for every other entry in the column is the difference between this underlined value and the other entry. If the weather is good and a medium quantity was bought, the regret is $10 - 7 = 3$. If the weather is good and a small quantity was bought, the regret is $10 - 4 = 6$, and so on.

- Put the regrets into a matrix (Table 13.6).

**Table 13.6**

| | | Event | | |
|---|---|---|---|---|
| | | Weather good | Weather average | Weather poor |
| Alternatives | Large quantity | 0 | 2 | 6 |
| | Medium quantity | 3 | 0 | 2 |
| | Small quantity | 6 | 5 | 0 |

- For each alternative find the highest regret:

Large quantity      Maximum of [0, 2, 6] = 6

Medium quantity    Maximum of [3, 0, 2] = 3

Small quantity      Maximum of [6, 5, 0] = 6

- Choose the alternative with the lowest of these maximum regrets. This is the medium quantity.

We have now described three decision criteria, but these are not the only ones, and we could have described many others. An ambitious organisation might aim for the highest profit and use a criterion based on the highest return (a 'maximax profit' criterion). Or it may try to balance the best and worst outcomes for each event and use a criterion based on:

$$\{\alpha \times \text{best outcome}\} + \{(1 - \alpha) \times \text{worst outcome}\}$$

where $\alpha$ is given a value between zero and one.

Different criteria will, of course, suggest different alternatives. Then you have to use the most relevant. If the decision makers are working as consultants they may use the Savage criterion; if the decision is made for a small company that cannot afford high losses, then Wald may be best; if there is really nothing to choose between the different events, Laplace may be useful.

It is difficult to go beyond these general guidelines, but you might notice one other feature. Both the Wald and Savage criteria recommend their decision based on one outcome – the worst for Wald and the one that leads to the highest regret for Savage – so their choice might be affected by a few outlying results. Perhaps the main benefit of decision criteria is not their ability to suggest good alternatives, but the way that they show the structure of a problem and allow informed discussion.

## In summary

**The Savage criterion considers regret, which is the difference between the best outcome for an event and the actual outcome. Then it uses the Wald criterion on the regrets. There are many other possible decision criteria.**

# 13.4 Decisions with risk

## 13.4.1 Expected values

In the previous section we looked at decision making with uncertainty, where a number of events could happen, but we did not know the relative likelihood of each. With decision making under risk there are again a number of events, but this time we know the probability of each one.

The usual way of solving problems with risk is to find the **expected value** for each alternative. This is the sum of the probability times the value of the outcome:

expected value = $\Sigma$ (probability $\times$ value of outcome)

The expected value is the average gain (or cost) we get if the decision is repeated a large number of times. It is not the value we get *every* time.

For decision making under risk:

● Calculate the expected value for each alternative.

● Choose the alternative with the best expected value (i.e. highest value for gains and lowest value for costs).

## we Worked example 13.5

What is the best alternative for the matrix of gains shown in Table 13.7?

**Table 13.7**

| | | | Events | | |
|---|---|---|---|---|---|
| | | 1<br>P = 0.1 | 2<br>P = 0.2 | 3<br>P = 0.6 | 4<br>P = 0.1 |
| | A | 10 | 7 | 5 | 9 |
| Alternatives | B | 3 | 20 | 2 | 10 |
| | C | 3 | 4 | 11 | 1 |
| | D | 8 | 4 | 2 | 16 |

## Solution

The expected value for each alternative is the sum of the probability times the value of the outcome:

Alternative A: $(0.1 \times 10) + (0.2 \times 7) + (0.6 \times 5) + (0.1 \times 9) = 6.3$

Alternative B: $(0.1 \times 3) + (0.2 \times 20) + (0.6 \times 2) + (0.1 \times 10) = 6.5$

Alternative C: $(0.1 \times 3) + (0.2 \times 4) + (0.6 \times 11) + (0.1 \times 1) = 7.8$

Alternative D: $(0.1 \times 8) + (0.2 \times 4) + (0.6 \times 2) + (0.1 \times 16) = 4.4$

As these are gains, the best alternative is C with an expected value of 7.8. If this decision is repeated a large number of times, the average return will be 7.8; if the decision is made only once, the gain can be any of the four values 3, 4, 11 or 1.

## we Worked example 13.6

A transport firm bids for a contract to move newspapers from a printing works to a wholesaler. It can make one of three tenders: a low one that assumes newspaper sales will rise and unit transport costs will decrease; a medium one that gives a reasonable return if newspaper sales stay the same; or a high one that assumes newspaper sales will fall and unit transport costs will increase. The probabilities of newspaper sales and profits (in thousands of pounds) for the transport firm are shown in Table 13.8. Which tender should it submit?

**Table 13.8**

| | Newspaper sales | | |
|---|---|---|---|
| | Decrease $P = 0.4$ | Stay same $P = 0.3$ | Increase $P = 0.3$ |
| Low tender | 10 | 15 | 16 |
| Medium tender | 5 | 20 | 10 |
| High tender | 18 | 10 | −5 |

## Solution

Calculating the expected value for each alternative:

Low tender: $(0.4 \times 10) + (0.3 \times 15) + (0.3 \times 16) = 13.3$

Medium tender: $(0.4 \times 5) + (0.3 \times 20) + (0.3 \times 10) = 11.0$

High tender: $(0.4 \times 18) + (0.3 \times 10) - (0.3 \times 5) = 8.7$

As these are profits the best alternative is the one with highest expected value, which is the low tender.

## In summary

**Expected values are defined as $\Sigma$ (probability $\times$ value of outcome). We can use these to suggest the best alternative in situations of risk.**

## 13.4.2 Using Bayes' theorem to update probabilities

In Chapter 10 we showed how Bayes' theorem can update conditional probabilities. Bayes' theorem states that:

$$P(a/b) = \frac{P(b/a) \times P(a)}{P(b)}$$

where:

$P(a/b)$ = probability of $a$ happening given that $b$ has already happened;

$P(b/a)$ = probability of $b$ given that $a$ has already happened;

$P(a)$, $P(b)$ = probabilities of $a$ and $b$ respectively.

We can use this result to find expected values for dependent events.

## we Worked example 13.7

The crowd for a sports event might be small (with a probability of 0.4) or large. The organisers can analyse advance sales of tickets a week before the event takes place. The advanced sales can be high, average or low, with the probability of advanced sales conditional on crowd size given by Table 13.9.

**Table 13.9**

| | | Advance sales | | |
|---|---|---|---|---|
| | | High | Average | Low |
| Crowd size | Large | 0.7 | 0.3 | 0.0 |
| | Small | 0.2 | 0.2 | 0.6 |

The organisers must chose one of two plans in running the event, and Table 13.10 below gives the net profit in thousands of pounds for each combination of plan and crowd size.

**Table 13.10**

| | | Plan 1 | Plan 2 |
|---|---|---|---|
| Crowd size | Large | 10 | 14 |
| | Small | 9 | 5 |

If the organisers use the information on advance sales, how can they maximise their expected profits? How much should they pay for the information on advance sales?

## Solution

We can use the following abbreviations:

CL and CS for crowd size large and crowd size small.

ASH, ASA and ASL for advance sales high, average and low.

If the organisers do not use the information on advance sales, the best they can do is use the probabilities of large and small crowds, 0.6 and 0.4 respectively. Then the expected values for the two plans are:

Plan 1: $(0.6 \times 10) + (0.4 \times 9) = 9.6$

Plan 2: $(0.6 \times 14) + (0.4 \times 5) = 10.4$      better plan

Hence they would use plan 2 with an expected value of £10 400.

The information on advance ticket sales gives the conditional probabilities $P(ASH/CL)$, $P(ASH/CS)$, etc. but the organisers want these the other way around, $P(CL/ASH)$, $P(CS/ASH)$, etc. We can use Bayes' theorem to find these, and a standard package gives the results shown in Figure 13.2.

|  | **** DATA ENTERED **** | | |
|---|---|---|---|
| PRIOR PROBABILITIES | P(CL) | P(CS) | |
| 1 Crowd size large, P(CL) | 0.60 | | |
| 2 Crowd size small, P(CS) | 0.40 | | |
| CONDITIONAL PROBABILITIES | P(ASH/CL) | P(ASA/CS) | etc. |
|  | 1 High, ASH | 2 Average, ASA | 3 Low, ASL |
| 1 Crowd size large, CL | 0.70 | 0.30 | 0.00 |
| 2 Crowd size small, CS | 0.20 | 0.20 | 0.60 |
|  | **** RESULTS CALCULATED **** | | |
| JOINT PROBABILITIES | P(CL and ASH) | P(CS and ASA) | etc. |
|  | 1 High, ASH | 2 Average, ASA | 3 Low, ASL |
| 1 Crowd size large, CL | 0.42 | 0.18 | 0.00 |
| 2 Crowd size small, CS | 0.08 | 0.08 | 0.24 |
| MARGINAL PROBABILITIES | P(ASH) | P(ASA) | P(ASL) |
| 1 High P(ASH) | 0.50 | | |
| 2 Average P(ASA) | 0.26 | | |
| 3 Low P(ASL) | 0.24 | | |
| RESULTS FOR BAYES' THEOREM | | | |
| Revised probabilities | P(CL/ASH) | P(CS/ASA) | etc. |
|  | 1 High, ASH | 2 Average, ASA | 3 Low, ASL |
| 1 Crowd size large, CL | 0.84 | 0.69 | 0.00 |
| 2 Crowd size small, CS | 0.16 | 0.31 | 1.00 |

**Figure 13.2** Printout for Bayes' theorem calculations for Worked Example 13.7.

If the advance sales are high, the probability of a large crowd is 0.84 and of a small crowd is 0.16.

- If the organisers choose plan 1, the expected value is

$$(0.84 \times 10) + (0.16 \times 9) = 9.84$$

- If the organisers choose plan 2, the expected value is

$$(0.84 \times 14) + (0.16 \times 5) = 12.56$$

So with advance sales high, they will choose plan 2.

We can use this reasoning for the other alternatives:

ASH: Plan 1  $(0.84 \times 10) + (0.16 \times 9) = 9.84$
      Plan 2  $(0.84 \times 14) + (0.16 \times 5) = 12.56$      better plan

ASA: Plan 1  $(0.69 \times 10) + (0.31 \times 9) = 9.69$
      Plan 2  $(0.69 \times 14) + (0.31 \times 5) = 11.21$      better plan

ASL: Plan 1  $(0.00 \times 10) + (1.00 \times 9) = 9.00$      better plan
      Plan 2  $(0.00 \times 14) + (1.00 \times 5) = 5.00$

So the best decisions for the organisers are:

- If the advance sales are high or average choose plan 2.

- If the advance sales are low choose plan 1.

We can go one step further than this, as we know the probability of high, average and low advance sales is, respectively, 0.5, 0.26 and 0.24. So we can calculate the overall expected value of following the recommended strategy as:

$$(0.5 \times 12.56) + (0.26 \times 11.21) + (0.24 \times 9) = 11.35$$

This profit of £11 350 compares with £10 400 when the advance sales information is not used. So the value of the additional information is $11\,350 - 10\,400 = £950$.

## In summary

**We can use Bayes' theorem to update probabilities and find a value of additional information.**

## 13.4.3 Utilities

Expected values are easy to use but they do not always reflect real preferences. Consider the payoff matrix of Table 13.11, which has a 90% chance of giving a loss.

**Table 13.11**

| | | Events | |
| --- | --- | --- | --- |
| | | Gain P = 0.1 | Loss P = 0.9 |
| Alternatives | Invest | £500 000 | −£50 000 |
| | Do not invest | £0 | £0 |

The expected values are:

Invest        $(0.1 \times 500\,000) - (0.9 \times 50\,000) = £5000$        better plan

Do not invest    $(0.1 \times 0) + (0.9 \times 0)$            $= £0$

Expected values suggest investing, even though there is a 90% chance of losing. The problem is that expected values give the average value when the decision is repeated a large number of times. However, if a decision is made only once, they can give misleading advice. They also assume a linear relationship between the amount of money and its value – thus £100 has a value one hundred times higher than £1. In practice this strict linear relationship is not true. We can use **utilities** to overcome these problems and to give a more accurate view of the value of money. Figure 13.3 shows a typical utility function.

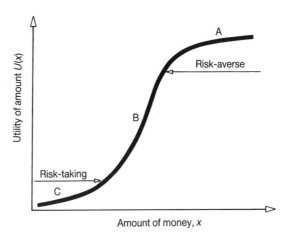

**Figure 13.3**    Utility curve relating amount of money to its value.

This utility function has three distinct regions. At the top, near point A, the utility is rising slowly with the amount of money. Decision makers in this region already have a lot of money and would not put a high value on even more. Nevertheless, they would certainly not like to lose money and move nearer to point B where the utility falls very quickly. Gaining an amount of money is not very attractive, but losing it is very unattractive – so this gives conservative decision makers who do not like to take risks.

Region B on the graph has the utility of money almost linear, which is the assumption of expected values. Decision makers at point C do not have much money, so losing some will have little effect on their utility. However, gaining money and moving nearer to B is very attractive. Decision makers here will be keen to make a gain, and not mind a loss, which is characteristic of a risk taker.

Although utilities are useful in principle, it is very difficult to find a reasonable function. Each individual and organisation puts a different value on money and will have a different utility function – and even these curves vary over time. It is so difficult to find a useful function, that utility theory is not widely used in practice. In principle, once we have found a utility function, the process of choosing the best alternative is the same as with expected values, but with expected utilities replacing expected values.

## we Worked example 13.8

A person's utility curve is a reasonable approximation to $\sqrt{x}$. What is their best decision with the gains matrix of Table 13.12?

**Table 13.12**

|            |   | Events        |               |               |
|------------|---|---------------|---------------|---------------|
|            |   | $X$<br>$P=0.7$ | $Y$<br>$P=0.2$ | $Z$<br>$P=0.1$ |
| Alternatives | A | 14 | 24 | 12 |
|            | B | 6  | 40 | 90 |
|            | C | 1  | 70 | 30 |
|            | D | 12 | 12 | 6  |

## Solution

The calculations are similar to those for expected values, but the amount of money, $\sqrt{x}$, is replaced by its utility, $U(x)$. In this case $U(x) = \sqrt{x}$.

Alternative A    $(0.7 \times \sqrt{14}) + (0.2 \times \sqrt{24}) + (0.1 \times \sqrt{12}) = 3.95$    best alternative

Alternative B    $(0.7 \times \sqrt{6}) + (0.2 \times \sqrt{40}) + (0.1 \times \sqrt{90}) = 3.93$

Alternative C    $(0.7 \times \sqrt{1}) + (0.2 \times \sqrt{70}) + (0.1 \times \sqrt{30}) = 2.92$

Alternative D    $(0.7 \times \sqrt{12}) + (0.2 \times \sqrt{12}) + (0.1 \times \sqrt{6}) = 3.36$

Although the difference is small, the best alternative is A. (If you calculate the results from expected values you will find that alternative B is the best.)

## In summary

**Utilities show the real value of money. We can use expected utilities to help decisions in situations of risk, provided we can find a realistic utility function.**

# 13.5 Decision trees

There are many situations where one decision leads to a series of other decisions. If you want to buy a car, your first decision might be to choose a new one or a secondhand one. If you choose a new car, you still have the choice of British, Japanese, French, German, Italian or others. If you choose a British car the choice is Rover, Jaguar, Ford, Rolls Royce, Vauxhall and so on. Then if you choose a Rover, you have to make a range of other decisions. Each decision you make opens a series of other choices (or sometimes events). We can represent these by a **decision tree**. This has the alternatives and events represented by the branches of a horizontal tree.

## we Worked example 13.9

A company asks a bank manager for a loan to finance an expansion. The bank manager has to decide whether or not to grant the loan. If the bank manager grants the loan, the company expansion may be successful or it may not. If the bank manager does not grant the loan, the company may continue banking as before or it may move its account to another bank. Draw a decision tree of this situation.

## Solution

Decision trees have a notional timescale going from left to right, with early decisions or events on the left, followed by later ones towards the right. There is only one decision in this example, followed by events over which the bank manager has no control, so the sequence is:

- the manager makes a decision;
- one of several possible events happens.

This is shown in the decision tree in Figure 13.4.
    The branches of the decision tree show the alternatives and events. There are three types of node, which are the points from which branches come.

○    **Random node** showing points where events happen, so all branches leaving random nodes have known probabilities.

☐    **Decision node** showing points where decisions are made, so all branches leaving a decision node are alternatives, and we choose the best.

│    **Terminal node** at the right-hand side of the tree and showing the ends of all sequences of decisions and events.

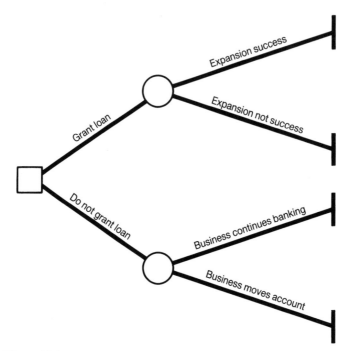

**Figure 13.4**   Decision tree for company expansion in Worked Example 13.9.

Now we have the basic structure of the tree, we can add probabilities and values. Suppose the bank currently values its business with the company at £2000 a year. If the manager grants the loan and the expansion succeeds, the value to the bank of increased business and interest charges is £3000 a year. If the expansion does not succeed, the bank will still have business valued at £1000 a year. There is a probability of 0.7 that the expansion plan will be successful. If the manager does not grant the loan there is a probability of 0.6 that the company will transfer its account to another bank.

Figure 13.5 shows the decision tree with these figures added. The probabilities are on the appropriate event branches – and if all events are included, the sum of the probabilities from each random node equals one. Values have been put on terminal nodes, showing the total value of moving through the tree and reaching the terminal node. In this case these values show the annual business expected by the bank.

The next part of the analysis moves from right to left through the tree and puts a value on each node. We do this by finding the best decision at each decision node and the expected value at each random node.

- At each decision node the alternative branches leaving are connected to following nodes. The values on these following nodes are compared, the best branch is chosen and the node value is transferred.

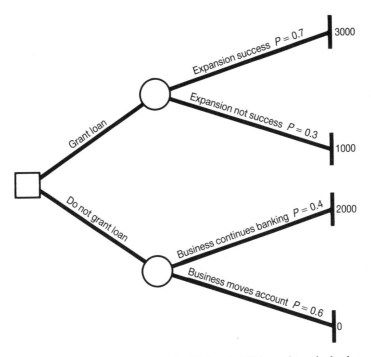

**Figure 13.5**   Decision tree with added probabilities and terminal values.

- The value at each random node is the expected value of the branches leaving. This is the sum, for all branches, of the probability of leaving by the branch multiplied by the value of the node at the end of the branch.

Adding these values to the tree in Figure 13.5 gives the results shown in Figure 13.6. The value at the left-hand, origin node is the overall expected value.
  The calculations are as follows:

- At random node 1 calculate the expected value:

$$(0.7 \times 3000) + (0.3 \times 1000) = 2400$$

- At random node 2 calculate the expected value:

$$(0.4 \times 2000) + (0.6 \times 0) = 800$$

- At decision node 3 choose the best alternative:

$$\text{Maximum of } [2400, 800] = 2400$$

The best policy is to grant the loan and this will have an expected value of £2400.

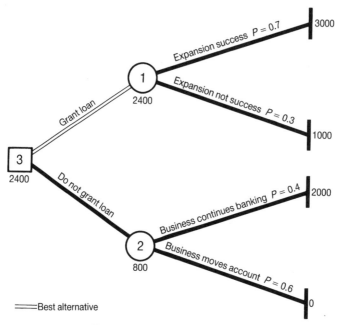

**Figure 13.6** Analysing a decision tree.

## we Worked example 13.10

A workshop is about to install a new machine for stamping and pressing parts for domestic appliances. Three suppliers have made bids to supply the machine.

The first supplier offers the Basicor machine which automatically produces parts of acceptable, but not outstanding quality. The output from the machine is variable (depending on material supplied and a variety of settings) and might be 1000 a week (with probability 0.1), 2000 a week (with probability 0.7) or 3000 a week. The profit for this machine is £4 a unit.

The second supplier offers a Superstamp machine that makes higher quality parts than the Basicor. The output from this might be 700 a week (with probability 0.4) or 1000 a week, with a notional profit of £10 a unit.

The third supplier offers the Switchover machine which can be set to produce either 1300 high quality parts a week at a profit of £6 a unit, or 1600 medium quality parts a week with a profit of £5 a unit.

If the chosen machine produces 2000 or more units a week, all production may be exported as a single bulk order. Then there is a 60% chance of selling for 50% more profit, and a 40% chance of selling for 50% less profit.

What is the expected profit?

## Solution

Figure 13.7 shows the tree for this decision. The terminal nodes give the weekly profit, which is the number produced multiplied by the profit per unit. If 1000 are produced on the Basicor machine the value is £4000, and so on. If the output from Basicor is exported, profit may rise by 50% (i.e. to £6 a unit) or fall by 50% (i.e. to £2 a unit).

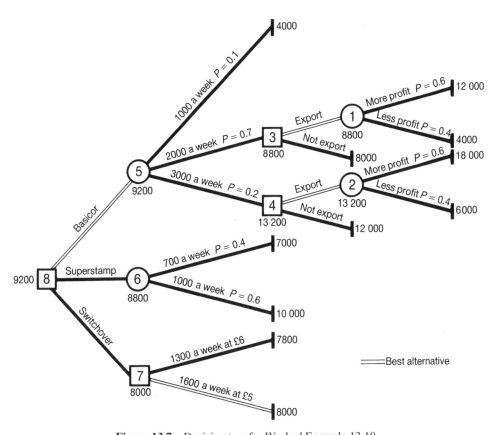

**Figure 13.7**   Decision tree for Worked Example 13.10.

Calculations at each node are as follows:

- At random node 1 calculate the expected value:

$$(0.6 \times 12\,000) + (0.4 \times 4000) = 8800$$

- At random node 2 calculate the expected value:

$$(0.6 \times 18\,000) + (0.4 \times 6000) = 13\,200$$

- At decision node 3 choose the best alternative:

$$\text{Max}[8800, 8000] = 8800$$

- At decision node 4 choose the best alternative:

$$\text{Max}[13\,200,\, 12\,000] = 13\,200$$

- At random node 5 calculate the expected value:

$$(0.1 \times 4000) + (0.7 \times 8800) + (0.2 \times 13\,200) = 9200$$

- At random node 6 calculate the expected value:

$$(0.4 \times 7000) + (0.6 \times 10\,000) = 8800$$

- At decision node 7 choose the best alternative:

$$\text{Max}[7800,\, 8000] = 8000$$

- At decision node 8 choose the best alternative:

$$\text{Max}[9200,\, 8800,\, 8000] = 9200$$

The best policy is to buy the Basicor machine and if it produces more than 2000 units, export all production. The expected profit from this policy is £9200 a week.

Decision trees can be draw by a computer, but most software gives tables of results without actually drawing a tree. Figure 13.8 shows a typical printout for Worked Example 13.10.

DATA ENTERED

MAXIMISATION PROBLEM

**** DECISION NODES ****

| NODE | NUMBER OF BRANCHES | ALTERNATIVE NUMBER | ENDING NODE |
|------|--------------------|--------------------|-------------|
| 3 | 2 | 1 Export | 1 |
|   |   | 2 Not export | 11 |
| 4 | 2 | 1 Export | 2 |
|   |   | 2 Not export | 15 |
| 7 | 2 | 1 1300 a week | 18 |
|   |   | 2 1600 a week | 19 |
| 8 | 3 | 1 Basicor | 5 |
|   |   | 2 Superstamp | 6 |
|   |   | 3 Switchover | 7 |

**Figure 13.8**  Analysis of decision tree for Worked Example 13.10.  (*continued*)

**** RANDOM NODES ****

| NODE | NUMBER OF BRANCHES | PROBABILITY | ENDING NODE |
|---|---|---|---|
| 1 | 2 | 1  0.6 More profit | 10 |
|   |   | 2  0.4 Less profit | 11 |
| 2 | 2 | 1  0.6 More profit | 13 |
|   |   | 2  0.4 Less profit | 14 |
| 5 | 3 | 1  0.1 1000 a week | 9 |
|   |   | 2  0.7 2000 a week | 3 |
|   |   | 3  0.2 3000 a week | 4 |
| 6 | 2 | 1  0.4 700 a week | 16 |
|   |   | 2  0.6 1000 a week | 17 |

**** TERMINAL NODES ****

| NODE | VALUE |
|---|---|
| 9 | 4000 |
| 10 | 12000 |
| 11 | 4000 |
| 12 | 8000 |
| 13 | 18000 |
| 14 | 6000 |
| 15 | 12000 |
| 16 | 7000 |
| 17 | 10000 |
| 18 | 7800 |
| 19 | 8000 |

RESULTS

MAXIMISATION PROBLEM

| NODE | VALUE | BEST ALTERNATIVE | |
|---|---|---|---|
| 1 | 8800 | | |
| 2 | 13200 | | |
| 3 | 8800 | 1 Export | |
| 4 | 13200 | 1 Export | |
| 5 | 9200 | | |
| 6 | 8800 | | |
| 7 | 8000 | 2 1600 a week | |
| 8 | 9200 | 1 Basicor | ********* Origin |
| 9 | 4000 | Terminal | |
| 10 | 12000 | Terminal | |
| 11 | 4000 | Terminal | |
| 12 | 8000 | Terminal | |
| 13 | 18000 | Terminal | |
| 14 | 6000 | Terminal | |
| 15 | 12000 | Terminal | |
| 16 | 7000 | Terminal | |
| 17 | 10000 | Terminal | |
| 18 | 78000 | Terminal | |
| 19 | 8000 | Terminal | |

OVERALL VALUE  9200

**Figure 13.8 (cont).**  Analysis of decision tree for Worked Example 13.10.

## In summary

We can show sequential decisions in a decision tree. For this we:

- determine the alternatives, events and their probabilities, outcomes, etc.;
- draw a tree (moving from left to right), showing the sequences of events and alternatives, and setting the values at terminal nodes;
- analyse the tree (moving from right to left), choosing the best alternative at decision nodes and calculating the expected value at random nodes.

## Chapter review

This chapter described some aspects of decision analysis. In particular, it:

- outlined the characteristics of decision making;
- described decision maps and payoff matrices for analysing decisions;
- mentioned decision making under certainty;
- used decision criteria for decision making with uncertainty;
- calculated expected values for decision making with risk;
- mentioned utilities;
- discussed sequential decisions and decision trees.

## Problems

13.1   A pub on the seafront at Blackpool notices that its profits are declining. The landlord has several ways of increasing profits – attracting more customers, increasing prices, getting customers to spend more, etc. – but each of these leads to a string of effects. Draw a map of the interactions for this problem.

13.2   Choose the best alternatives in the matrix of gains given in Table 13.13.

**Table 13.13**

|  |  | Event |
|---|---|---|
| Alternatives | a | 100 |
|  | b | 950 |
|  | c | −250 |
|  | d | 0 |
|  | e | 950 |
|  | f | 500 |

13.3 Use the Laplace, Wald and Savage decision criteria to choose alternatives in the following matrices.

(a) Cost matrix

|  |  | Events | | | | |
|---|---|---|---|---|---|---|
|  |  | 1 | 2 | 3 | 4 | 5 |
| Alternatives | a | 100 | 70 | 115 | 95 | 60 |
|  | b | 95 | 120 | 120 | 90 | 150 |
|  | c | 180 | 130 | 60 | 160 | 120 |
|  | d | 80 | 75 | 50 | 100 | 95 |
|  | e | 60 | 140 | 100 | 170 | 160 |

(b) Gains matrix

|  |  | Events | | | |
|---|---|---|---|---|---|
|  |  | 1 | 2 | 3 | 4 |
| Alternatives | a | 1 | 6 | 3 | 7 |
|  | b | 2 | 5 | 1 | 4 |
|  | c | 8 | 1 | 4 | 2 |
|  | d | 5 | 2 | 7 | 8 |

13.4 (a) What is the best alternative in the matrix of gains of Table 13.14?

**Table 13.14**

|  |  | Events | | |
|---|---|---|---|---|
|  |  | 1<br>$P=0.4$ | 2<br>$P=0.3$ | 3<br>$P=0.3$ |
| Alternatives | a | 100 | 90 | 120 |
|  | b | 80 | 102 | 110 |

(b) Would this decision change if a utility function $U(x) = \sqrt{x}$ is used?

13.5 A company can launch one of three versions of a new product, X, Y or Z. The profit depends on market reaction and there is a 30% chance that this will be good, a 40% chance it will be medium and a 30% chance it will be poor. Which version should the company launch if profits are as given in Table 13.15?

**Table 13.15**

|  |  | Market reaction | | |
|---|---|---|---|---|
|  |  | Good | Medium | Poor |
| Version | X | 100 | 110 | 80 |
|  | Y | 70 | 90 | 120 |
|  | Z | 130 | 100 | 70 |

A survey can give more information on market reaction. Experience suggests that a survey will give results A, B or C with probabilities $P(A/Good)$, etc. shown in the Table 13.16.

**Table 13.16**

|  |  | Result | | |
|---|---|---|---|---|
|  |  | A | B | C |
| Market reaction | Good | 0.2 | 0.2 | 0.6 |
|  | Medium | 0.2 | 0.5 | 0.3 |
|  | Poor | 0.4 | 0.3 | 0.3 |

How much should the company pay for this information?

13.6 A road haulage contractor owns a lorry with a one-year-old engine. He has to decide now, and again in one year's time, whether or not to replace the engine. If he replaces it, the cost is £500. If he does not replace it there is an increased chance it will break down during the year and the cost of replacing an engine then is £800. If an engine is replaced during the year, the replacement engine is assumed to be one year old at the time when the next decision is taken. The probability of breakdown of an engine during a year is:

|  | Age of engine (years) | | |
|---|---|---|---|
|  | 0 | 1 | 2 |
| Probability of breakdown | 0.0 | 0.2 | 0.7 |

Draw a decision tree for this problem and find the decisions that minimise the cost over the next two years.

13.7 An organisation is considering launching an entirely new service. If the market reaction to this service is good (which has a probability of 0.2) the organisation will make £3000 a week; if market reaction is medium (with probability 0.5) it will make £1000 a week, but if reaction is poor (with probability 0.3) it will lose £1500 a week. The organisation can run a survey to test market reaction with results A, B or C. The reliability of these surveys is described by the following matrix of $P(A/good)$, etc. Use a decision tree to find how much the organisation should pay for this survey.

|  |  | Result | | |
|---|---|---|---|---|
|  |  | A | B | C |
| Market reaction | Good | 0.7 | 0.2 | 0.1 |
|  | Medium | 0.2 | 0.6 | 0.2 |
|  | Poor | 0.1 | 0.4 | 0.5 |

# CS Case study – The Newisham Reservoir

Newisham has a population of about 30 000. It used to get its water supply from the River Feltham, but industrial developments upstream began diverting more of the water. When the flow reaching the Newisham water treatment works became too small, the town decided to build a reservoir by damming the Feltham and diverting tributaries. Unfortunately, the dam reduced the amount of water available to farmers downstream.

One farmer finds the water supply to his cattle has all but dried up. He now has the option of either connecting to the local mains water supply at a cost of £44 000 or drilling a new well. The cost of the well is not known with certainty but could be £32 000 (with a probability of 0.3), £44 000 (with a probability of 0.3) or £56 000, depending on the underground rock structure and depth of water.

A local water survey company can do on-site tests. For a cost of £600 it will give either a favourable or an unfavourable report on the chances of easily finding water. The reliability of this report (phrased in terms of the probability of a favourable report given the drilling cost will be low, etc.) is given in the following table. What should the farmer do?

| | Drilling well cost | | |
| --- | --- | --- | --- |
| | £32 000 | £44 000 | £56 000 |
| Favourable report | 0.8 | 0.6 | 0.2 |
| Unfavourable report | 0.2 | 0.4 | 0.8 |

# Appendix A

# Probabilities for the binomial distribution

| | | | | | | $p$ | | | | | |
|---|---|---|---|---|---|---|---|---|---|---|---|
| $n$ | $r$ | .05 | .10 | .15 | .20 | .25 | .30 | .35 | .40 | .45 | .50 |
| 1 | 0 | .9500 | .9000 | .8500 | .8000 | .7500 | .7000 | .6500 | .6000 | .5500 | .5000 |
|   | 1 | .0500 | .1000 | .1500 | .2000 | .2500 | .3000 | .3500 | .4000 | .4500 | .5000 |
| 2 | 0 | .9025 | .8100 | .7225 | .6400 | .5625 | .4900 | .4225 | .3600 | .3025 | .2500 |
|   | 1 | .0950 | .1800 | .2550 | .3200 | .3750 | .4200 | .4550 | .4800 | .4950 | .5000 |
|   | 2 | .0025 | .0100 | .0225 | .0400 | .0625 | .0900 | .1225 | .1600 | .2025 | .2500 |
| 3 | 0 | .8574 | .7290 | .6141 | .5120 | .4219 | .3430 | .2746 | .2160 | .1664 | .1250 |
|   | 1 | .1354 | .2430 | .3251 | .3840 | .4219 | .4410 | .4436 | .4320 | .4084 | .3750 |
|   | 2 | .0071 | .0270 | .0574 | .0960 | .1406 | .1890 | .2389 | .2880 | .3341 | .3750 |
|   | 3 | .0001 | .0010 | .0034 | .0080 | .0156 | .0270 | .0429 | .0640 | .0911 | .1250 |
| 4 | 0 | .8145 | .6561 | .5220 | .4096 | .3164 | .2401 | .1785 | .1296 | .0915 | .0625 |
|   | 1 | .1715 | .2916 | .3685 | .4096 | .4219 | .4116 | .3845 | .3456 | .2995 | .2500 |
|   | 2 | .0135 | .0486 | .0975 | .1536 | .2109 | .2646 | .3105 | .3456 | .3675 | .3750 |
|   | 3 | .0005 | .0036 | .0115 | .0256 | .0469 | .0756 | .1115 | .1536 | .2005 | .2500 |
|   | 4 | .0000 | .0001 | .0005 | .0016 | .0039 | .0081 | .0150 | .0256 | .0410 | .0625 |
| 5 | 0 | .7738 | .5905 | .4437 | .3277 | .2373 | .1681 | .1160 | .0778 | .0503 | .0312 |
|   | 1 | .2036 | .3280 | .3915 | .4096 | .3955 | .3602 | .3124 | .2592 | .2059 | .1562 |
|   | 2 | .0214 | .0729 | .1382 | .2048 | .2637 | .3087 | .3364 | .3456 | .3369 | .3125 |
|   | 3 | .0011 | .0081 | .0244 | .0512 | .0879 | .1323 | .1811 | .2304 | .2757 | .3125 |
|   | 4 | .0000 | .0004 | .0022 | .0064 | .0146 | .0284 | .0488 | .0768 | .1128 | .1562 |
|   | 5 | .0000 | .0000 | .0001 | .0003 | .0010 | .0024 | .0053 | .0102 | .0185 | .0312 |
| 6 | 0 | .7351 | .5314 | .3771 | .2621 | .1780 | .1176 | .0754 | .0467 | .0277 | .0156 |
|   | 1 | .2321 | .3543 | .3993 | .3932 | .3560 | .3025 | .2437 | .1866 | .1359 | .0938 |
|   | 2 | .0305 | .0984 | .1762 | .2458 | .2966 | .3241 | .3280 | .3110 | .2780 | .2344 |
|   | 3 | .0021 | .0146 | .0415 | .0819 | .1318 | .1852 | .2355 | .2765 | .3032 | .3125 |
|   | 4 | .0001 | .0012 | .0055 | .0154 | .0330 | .0595 | .0951 | .1382 | .1861 | .2344 |
|   | 5 | .0000 | .0001 | .0004 | .0015 | .0044 | .0102 | .0205 | .0369 | .0609 | .0938 |
|   | 6 | .0000 | .0000 | .0000 | .0001 | .0002 | .0007 | .0018 | .0041 | .0083 | .0516 |

| | | | | | | $p$ | | | | | |
|---|---|---|---|---|---|---|---|---|---|---|---|
| $n$ | $r$ | .05 | .10 | .15 | .20 | .25 | .30 | .35 | .40 | .45 | .50 |
| 7 | 0 | .6983 | .4783 | .3206 | .2097 | .1335 | .0824 | .0490 | .0280 | .0152 | .0078 |
| | 1 | .2573 | .3720 | .3960 | .3670 | .3115 | .2471 | .1848 | .1306 | .0872 | .0547 |
| | 2 | .0406 | .1240 | .2097 | .2753 | .3115 | .3177 | .2985 | .2613 | .2140 | .1641 |
| | 3 | .0036 | .0230 | .0617 | .1147 | .1730 | .2269 | .2679 | .2903 | .2918 | .2734 |
| | 4 | .0002 | .0026 | .0109 | .0287 | .0577 | .0972 | .1442 | .1935 | .2388 | .2734 |
| | 5 | .0009 | .0002 | .0012 | .0043 | .0115 | .0250 | .0466 | .0774 | .1172 | .1641 |
| | 6 | .0000 | .0000 | .0001 | .0004 | .0013 | .0036 | .0084 | .0172 | .0320 | .0547 |
| | 7 | .0000 | .0000 | .0000 | .0000 | .0001 | .0002 | .0006 | .0016 | .0037 | .0078 |
| 8 | 0 | .6634 | .4305 | .2725 | .1678 | .1001 | .0576 | .0319 | .0168 | .0084 | .0039 |
| | 1 | .2793 | .3826 | .3847 | .3355 | .2670 | .1977 | .1373 | .0896 | .0548 | .0312 |
| | 2 | .0515 | .1488 | .2376 | .2936 | .3115 | .2965 | .2587 | .2090 | .1569 | .1094 |
| | 3 | .0054 | .0331 | .0839 | .1468 | .2076 | .2541 | .2786 | .2787 | .2568 | .2188 |
| | 4 | .0004 | .0046 | .0185 | .0459 | .0865 | .1361 | .1875 | .2322 | .2627 | .2734 |
| | 5 | .0000 | .0004 | .0026 | .0092 | .0231 | .0467 | .0808 | .1239 | .1719 | .2188 |
| | 6 | .0000 | .0000 | .0002 | .0011 | .0038 | .0100 | .0217 | .0413 | .0703 | .1094 |
| | 7 | .0000 | .0000 | .0000 | .0001 | .0004 | .0012 | .0033 | .0079 | .0164 | .0312 |
| | 8 | .0000 | .0000 | .0000 | .0000 | .0000 | .0001 | .0002 | .0007 | .0017 | .0039 |
| 9 | 0 | .6302 | .3874 | .2316 | .1342 | .0751 | .0404 | .0207 | .0101 | .0046 | .0020 |
| | 1 | .2985 | .3874 | .3679 | .3020 | .2253 | .1556 | .1004 | .0605 | .0339 | .0176 |
| | 2 | .0629 | .1722 | .2597 | .3020 | .3003 | .2668 | .2162 | .1612 | .1110 | .0703 |
| | 3 | .0077 | .0446 | .1069 | .1762 | .2336 | .2668 | .2716 | .2508 | .2119 | .1641 |
| | 4 | .0006 | .0074 | .0283 | .0661 | .1168 | .1715 | .2194 | .2508 | .2600 | .2461 |
| | 5 | .0000 | .0008 | .0050 | .0165 | .0389 | .0735 | .1181 | .1672 | .2128 | .2461 |
| | 6 | .0000 | .0001 | .0006 | .0028 | .0087 | .0210 | .0424 | .0743 | .1160 | .1641 |
| | 7 | .0000 | .0000 | .0000 | .0003 | .0012 | .0039 | .0098 | .0212 | .0407 | .0703 |
| | 8 | .0000 | .0000 | .0000 | .0000 | .0001 | .0004 | .0013 | .0035 | .0083 | .0716 |
| | 9 | .0000 | .0000 | .0000 | .0000 | .0000 | .0000 | .0001 | .0003 | .0008 | .0020 |
| 10 | 0 | .5987 | .3487 | .1969 | .1074 | .0563 | .0282 | .0135 | .0060 | .0025 | .0010 |
| | 1 | .3151 | .3874 | .3474 | .2684 | .1877 | .1211 | .0725 | .0403 | .0207 | .0098 |
| | 2 | .0746 | .1937 | .2759 | .3020 | .2816 | .2335 | .1757 | .1209 | .0763 | .0439 |
| | 3 | .0105 | .0574 | .1298 | .2013 | .2503 | .2668 | .2522 | .2150 | .1665 | .1172 |
| | 4 | .0010 | .0112 | .0401 | .0881 | .1460 | .2001 | .2377 | .2508 | .2384 | .2051 |
| | 5 | .0001 | .0015 | .0085 | .0264 | .0584 | .1029 | .1563 | .2007 | .2340 | .2461 |
| | 6 | .0000 | .0001 | .0012 | .0055 | .0162 | .0368 | .0689 | .1115 | .1596 | .2051 |
| | 7 | .0000 | .0000 | .0001 | .0008 | .0031 | .0090 | .0212 | .0425 | .0746 | .1172 |
| | 8 | .0000 | .0000 | .0000 | .0001 | .0004 | .0014 | .0043 | .0106 | .0229 | .0439 |
| | 9 | .0000 | .0000 | .0000 | .0000 | .0000 | .0001 | .0005 | .0016 | .0042 | .0098 |
| | 10 | .0000 | .0000 | .0000 | .0000 | .0000 | .0000 | .0000 | .0001 | .0003 | .0010 |

| | | | | | | $p$ | | | | | |
|---|---|---|---|---|---|---|---|---|---|---|---|
| $n$ | $r$ | .05 | .10 | .15 | .20 | .25 | .30 | .35 | .40 | .45 | .50 |
| 11 | 0 | .5688 | .3138 | .1673 | .0859 | .0422 | .0198 | .0088 | .0036 | .0014 | .0005 |
| | 1 | .3293 | .3835 | .3248 | .2362 | .1549 | .0932 | .0518 | .0266 | .0125 | .0054 |
| | 2 | .0867 | .2131 | .2866 | .2953 | .2581 | .1998 | .1395 | .0887 | .0513 | .0269 |
| | 3 | .0137 | .0710 | .1517 | .2215 | .2581 | .2568 | .2254 | .1774 | .1259 | .0806 |
| | 4 | .0014 | .0158 | .0536 | .1107 | .1721 | .2201 | .2428 | .2365 | .2060 | .1611 |
| | 5 | .0001 | .0025 | .0132 | .0388 | .0803 | .1321 | .1830 | .2207 | .2360 | .2256 |
| | 6 | .0000 | .0003 | .0023 | .0097 | .0268 | .0566 | .0985 | .1471 | .1931 | .2256 |
| | 7 | .0000 | .0000 | .0003 | .0017 | .0064 | .0173 | .0379 | .0701 | .1128 | .1611 |
| | 8 | .0000 | .0000 | .0000 | .0002 | .0011 | .0037 | .0102 | .0234 | .0462 | .0806 |
| | 9 | .0000 | .0000 | .0000 | .0000 | .0001 | .0005 | .0018 | .0052 | .0126 | .0269 |
| | 10 | .0000 | .0000 | .0000 | .0000 | .0000 | .0000 | .0002 | .0007 | .0021 | .0054 |
| | 11 | .0000 | .0000 | .0000 | .0000 | .0000 | .0000 | .0000 | .0000 | .0002 | .0005 |
| 12 | 0 | .5404 | .2824 | .1422 | .0687 | .0317 | .0138 | .0057 | .0022 | .0008 | .0002 |
| | 1 | .3413 | .3766 | .3012 | .2062 | .1267 | .0712 | .0368 | .0174 | .0075 | .0029 |
| | 2 | .0988 | .2301 | .2924 | .2835 | .2323 | .1678 | .1088 | .0639 | .0339 | .0161 |
| | 3 | .0173 | .0852 | .1720 | .2362 | .2581 | .2397 | .1954 | .1419 | .0923 | .0537 |
| | 4 | .0021 | .0213 | .0683 | .1329 | .1936 | .2311 | .2367 | .2128 | .1700 | .1208 |
| | 5 | .0002 | .0038 | .0193 | .0532 | .1032 | .1585 | .2039 | .2270 | .2225 | .1934 |
| | 6 | .0000 | .0005 | .0040 | .0155 | .0401 | .0792 | .1281 | .1766 | .2124 | .2256 |
| | 7 | .0000 | .0000 | .0006 | .0033 | .0115 | .0291 | .0591 | .1009 | .1489 | .1934 |
| | 8 | .0000 | .0000 | .0001 | .0005 | .0024 | .0078 | .0199 | .0420 | .0762 | .1208 |
| | 9 | .0000 | .0000 | .0000 | .0001 | .0004 | .0015 | .0048 | .0125 | .0277 | .0537 |
| | 10 | .0000 | .0000 | .0000 | .0000 | .0000 | .0002 | .0008 | .0025 | .0068 | .0161 |
| | 11 | .0000 | .0000 | .0000 | .0000 | .0000 | .0000 | .0001 | .0003 | .0010 | .0029 |
| | 12 | .0000 | .0000 | .0000 | .0000 | .0000 | .0000 | .0000 | .0000 | .0001 | .0002 |
| 13 | 0 | .5133 | .2542 | .1209 | .0550 | .0238 | .0097 | .0037 | .0013 | .0004 | .0001 |
| | 1 | .3512 | .3672 | .2774 | .1787 | .1029 | .0540 | .0259 | .0113 | .0045 | .0016 |
| | 2 | .1109 | .2448 | .2937 | .2680 | .2059 | .1388 | .0836 | .0453 | .0220 | .0095 |
| | 3 | .0214 | .0997 | .1900 | .2457 | .2517 | .2181 | .1651 | .1107 | .0660 | .0349 |
| | 4 | .0028 | .0277 | .0838 | .1535 | .2097 | .2337 | .2222 | .1845 | .1350 | .0873 |
| | 5 | .0003 | .0055 | .0266 | .0691 | .1258 | .1803 | .2154 | .2214 | .1989 | .1571 |
| | 6 | .0000 | .0008 | .0063 | .0230 | .0559 | .1030 | .1546 | .1968 | .2169 | .2095 |
| | 7 | .0000 | .0001 | .0011 | .0058 | .0186 | .0442 | .0833 | .1312 | .1775 | .2095 |
| | 8 | .0000 | .0000 | .0001 | .0011 | .0047 | .0142 | .0336 | .0656 | .1089 | .1571 |
| | 9 | .0000 | .0000 | .0000 | .0001 | .0009 | .0034 | .0101 | .0243 | .0495 | .0873 |
| | 10 | .0000 | .0000 | .0000 | .0000 | .0001 | .0006 | .0022 | .0065 | .0162 | .0349 |
| | 11 | .0000 | .0000 | .0000 | .0000 | .0000 | .0001 | .0003 | .0012 | .0036 | .0095 |
| | 12 | .0000 | .0000 | .0000 | .0000 | .0000 | .0000 | .0000 | .0001 | .0005 | .0016 |
| | 13 | .0000 | .0000 | .0000 | .0000 | .0000 | .0000 | .0000 | .0000 | .0000 | .0001 |

| | | | | | | $p$ | | | | | |
|---|---|---|---|---|---|---|---|---|---|---|---|
| $n$ | $r$ | .05 | .10 | .15 | .20 | .25 | .30 | .35 | .40 | .45 | .50 |
| 14 | 0 | .4877 | .2288 | .1028 | .0440 | .0178 | .0068 | .0024 | .0008 | .0002 | .0001 |
| | 1 | .3593 | .3559 | .2539 | .1539 | .0832 | .0407 | .0181 | .0073 | .0027 | .0009 |
| | 2 | .1229 | .2570 | .2912 | .2501 | .1802 | .1134 | .0634 | .0317 | .0141 | .0056 |
| | 3 | .0259 | .1142 | .2056 | .2501 | .2402 | .1943 | .1366 | .0845 | .0462 | .0222 |
| | 4 | .0037 | .0348 | .0998 | .1720 | .2202 | .2290 | .2022 | .1549 | .1040 | .0611 |
| | 5 | .0004 | .0078 | .0352 | .0860 | .1468 | .1963 | .2178 | .2066 | .1701 | .1222 |
| | 6 | .0000 | .0013 | .0093 | .0322 | .0734 | .1262 | .1759 | .2066 | .2088 | .1833 |
| | 7 | .0000 | .0002 | .0019 | .0092 | .0280 | .0618 | .1082 | .1574 | .1952 | .2095 |
| | 8 | .0000 | .0000 | .0003 | .0020 | .0082 | .0232 | .0510 | .0918 | .1398 | .1833 |
| | 9 | .0000 | .0000 | .0000 | .0003 | .0018 | .0066 | .0183 | .0408 | .0762 | .1222 |
| | 10 | .0000 | .0000 | .0000 | .0000 | .0003 | .0014 | .0049 | .0136 | .0312 | .0611 |
| | 11 | .0000 | .0000 | .0000 | .0000 | .0000 | .0002 | .0010 | .0033 | .0093 | .0222 |
| | 12 | .0000 | .0000 | .0000 | .0000 | .0000 | .0000 | .0001 | .0005 | .0019 | .0056 |
| | 13 | .0000 | .0000 | .0000 | .0000 | .0000 | .0000 | .0000 | .0001 | .0002 | .0009 |
| | 14 | .0000 | .0000 | .0000 | .0000 | .0000 | .0000 | .0000 | .0000 | .0000 | .0001 |
| 15 | 0 | .4633 | .2059 | .0874 | .0352 | .0134 | .0047 | .0016 | .0005 | .0001 | .0000 |
| | 1 | .3658 | .3432 | .2312 | .1319 | .0668 | .0305 | .0126 | .0047 | .0016 | .0005 |
| | 2 | .1348 | .2669 | .2856 | .2309 | .1559 | .0916 | .0476 | .0219 | .0090 | .0032 |
| | 3 | .0307 | .1285 | .2184 | .2501 | .2252 | .1700 | .1110 | .0634 | .0318 | .0139 |
| | 4 | .0049 | .0428 | .1156 | .1876 | .2252 | .2186 | .1792 | .1268 | .0780 | .0417 |
| | 5 | .0006 | .0105 | .0449 | .1032 | .1651 | .2061 | .2123 | .1859 | .1404 | .0916 |
| | 6 | .0000 | .0019 | .0132 | .0430 | .0917 | .1472 | .1906 | .2066 | .1914 | .1527 |
| | 7 | .0000 | .0003 | .0030 | .0138 | .0393 | .0811 | .1319 | .1771 | .2013 | .1964 |
| | 8 | .0000 | .0000 | .0005 | .0035 | .0131 | .0348 | .0710 | .1181 | .1647 | .1964 |
| | 9 | .0000 | .0000 | .0001 | .0007 | .0034 | .0116 | .0298 | .0612 | .1048 | .1527 |
| | 10 | .0000 | .0000 | .0000 | .0001 | .0007 | .0030 | .0096 | .0245 | .0515 | .0916 |
| | 11 | .0000 | .0000 | .0000 | .0000 | .0001 | .0006 | .0024 | .0074 | .0191 | .0417 |
| | 12 | .0000 | .0000 | .0000 | .0000 | .0000 | .0001 | .0004 | .0016 | .0052 | .0139 |
| | 13 | .0000 | .0000 | .0000 | .0000 | .0000 | .0000 | .0001 | .0003 | .0010 | .0032 |
| | 14 | .0000 | .0000 | .0000 | .0000 | .0000 | .0000 | .0000 | .0000 | .0001 | .0005 |
| | 15 | .0000 | .0000 | .0000 | .0000 | .0000 | .0000 | .0000 | .0000 | .0000 | .0000 |

| $n$ | $r$ | .05 | .10 | .15 | .20 | .25 | $p$<br>.30 | .35 | .40 | .45 | .50 |
|---|---|---|---|---|---|---|---|---|---|---|---|
| 16 | 0 | .4401 | .1853 | .0743 | .0281 | .0100 | .0033 | .0010 | .0003 | .0001 | .0000 |
|  | 1 | .3706 | .3294 | .2097 | .1126 | .0535 | .0228 | .0087 | .0030 | .0009 | .0002 |
|  | 2 | .1463 | .2745 | .2775 | .2111 | .1336 | .0732 | .0353 | .0150 | .0056 | .0018 |
|  | 3 | .0359 | .1423 | .2285 | .2463 | .2079 | .1465 | .0888 | .0468 | .0215 | .0085 |
|  | 4 | .0061 | .0514 | .1311 | .2001 | .2252 | .2040 | .1553 | .1014 | .0572 | .0278 |
|  | 5 | .0008 | .0137 | .0555 | .1201 | .1802 | .2099 | .2008 | .1623 | .1123 | .0667 |
|  | 6 | .0001 | .0028 | .0180 | .0550 | .1101 | .1649 | .1982 | .1983 | .1684 | .1222 |
|  | 7 | .0000 | .0004 | .0045 | .0197 | .0524 | .1010 | .1524 | .1889 | .1969 | .1746 |
|  | 8 | .0000 | .0001 | .0009 | .0055 | .0197 | .0487 | .0923 | .1417 | .1812 | .1964 |
|  | 9 | .0000 | .0000 | .0001 | .0012 | .0058 | .0185 | .0442 | .0840 | .1318 | .1746 |
|  | 10 | .0000 | .0000 | .0000 | .0002 | .0014 | .0056 | .0167 | .0392 | .0755 | .1222 |
|  | 11 | .0000 | .0000 | .0000 | .0000 | .0002 | .0013 | .0049 | .0142 | .0337 | .0667 |
|  | 12 | .0000 | .0000 | .0000 | .0000 | .0000 | .0002 | .0011 | .0040 | .0115 | .0278 |
|  | 13 | .0000 | .0000 | .0000 | .0000 | .0000 | .0000 | .0002 | .0008 | .0029 | .0085 |
|  | 14 | .0000 | .0000 | .0000 | .0000 | .0000 | .0000 | .0000 | .0001 | .0005 | .0018 |
|  | 15 | .0000 | .0000 | .0000 | .0000 | .0000 | .0000 | .0000 | .0000 | .0001 | .0002 |
|  | 16 | .0000 | .0000 | .0000 | .0000 | .0000 | .0000 | .0000 | .0000 | .0000 | .0000 |
| 17 | 0 | .4181 | .1668 | .0631 | .0225 | .0075 | .0023 | .0007 | .0002 | .0000 | .0000 |
|  | 1 | .3741 | .3150 | .1893 | .0957 | .0426 | .0169 | .0060 | .0019 | .0005 | .0001 |
|  | 2 | .1575 | .2800 | .2673 | .1914 | .1136 | .0581 | .0260 | .0102 | .0035 | .0010 |
|  | 3 | .0415 | .1556 | .2359 | .2393 | .1893 | .1245 | .0701 | .0341 | .0144 | .0052 |
|  | 4 | .0076 | .0605 | .1457 | .2093 | .2209 | .1868 | .1320 | .0796 | .0411 | .0182 |
|  | 5 | .0010 | .0175 | .0668 | .1361 | .1914 | .2081 | .1849 | .1379 | .0875 | .0472 |
|  | 6 | .0001 | .0039 | .0236 | .0680 | .1276 | .1784 | .1991 | .1839 | .1432 | .0944 |
|  | 7 | .0000 | .0007 | .0065 | .0267 | .0668 | .1201 | .1685 | .1927 | .1841 | .1484 |
|  | 8 | .0000 | .0001 | .0014 | .0084 | .0279 | .0644 | .1134 | .1606 | .1883 | .1855 |
|  | 9 | .0000 | .0000 | .0003 | .0021 | .0093 | .0276 | .0611 | .1070 | .1540 | .1855 |
|  | 10 | .0000 | .0000 | .0000 | .0004 | .0025 | .0095 | .0263 | .0571 | .1008 | .1484 |
|  | 11 | .0000 | .0000 | .0000 | .0001 | .0005 | .0026 | .0090 | .0242 | .0525 | .0944 |
|  | 12 | .0000 | .0000 | .0000 | .0000 | .0001 | .0006 | .0024 | .0081 | .0215 | .0472 |
|  | 13 | .0000 | .0000 | .0000 | .0000 | .0000 | .0001 | .0005 | .0021 | .0068 | .0182 |
|  | 14 | .0000 | .0000 | .0000 | .0000 | .0000 | .0000 | .0001 | .0004 | .0016 | .0052 |
|  | 15 | .0000 | .0000 | .0000 | .0000 | .0000 | .0000 | .0000 | .0001 | .0003 | .0010 |
|  | 16 | .0000 | .0000 | .0000 | .0000 | .0000 | .0000 | .0000 | .0000 | .0000 | .0001 |
|  | 17 | .0000 | .0000 | .0000 | .0000 | .0000 | .0000 | .0000 | .0000 | .0000 | .0000 |

# Appendix
# B Probabilities for the Poisson distribution

| $r$ | $\mu$ |||||||||
|---|---|---|---|---|---|---|---|---|---|---|
| | .005 | .01 | .02 | .03 | .04 | .05 | .06 | .07 | .08 | .09 |
| 0 | .9950 | .9900 | .9802 | .9704 | .9608 | .9512 | .9418 | .9324 | .9231 | .9139 |
| 1 | .0050 | .0099 | .0192 | .0291 | .0384 | .0476 | .0565 | .0653 | .0738 | .0823 |
| 2 | .0000 | .0000 | .0002 | .0004 | .0008 | .0012 | .0017 | .0023 | .0030 | .0037 |
| 3 | .0000 | .0000 | .0000 | .0000 | .0000 | .0000 | .0000 | .0001 | .0001 | .0001 |

| $r$ | $\mu$ |||||||||
|---|---|---|---|---|---|---|---|---|---|---|
| | 0.1 | 0.2 | 0.3 | 0.4 | 0.5 | 0.6 | 0.7 | 0.8 | 0.9 | 1.0 |
| 0 | .9048 | .8187 | .7408 | .6703 | .6065 | .5488 | .4966 | .4493 | .4066 | .3679 |
| 1 | .0905 | .1637 | .2222 | .2681 | .3033 | .3293 | .3476 | .3595 | .3659 | .3679 |
| 2 | .0045 | .0164 | .0333 | .0536 | .0758 | .0988 | .1217 | .1438 | .1647 | .1839 |
| 3 | .0002 | .0011 | .0033 | .0072 | .0126 | .0198 | .0284 | .0383 | .0494 | .0613 |
| 4 | .0000 | .0001 | .0002 | .0007 | .0016 | .0030 | .0050 | .0077 | .0111 | .0153 |
| 5 | .0000 | .0000 | .0000 | .0001 | .0002 | .0004 | .0007 | .0012 | .0020 | .0031 |
| 6 | .0000 | .0000 | .0000 | .0000 | .0000 | .0000 | .0001 | .0002 | .0003 | .0005 |
| 7 | .0000 | .0000 | .0000 | .0000 | .0000 | .0000 | .0000 | .0000 | .0000 | .0001 |

| $r$ | $\mu$ |||||||||
|---|---|---|---|---|---|---|---|---|---|---|
| | 1.1 | 1.2 | 1.3 | 1.4 | 1.5 | 1.6 | 1.7 | 1.8 | 1.9 | 2.0 |
| 0 | .3329 | .3012 | .2725 | .2466 | .2231 | .2019 | .1827 | .1653 | .1496 | .1353 |
| 1 | .3662 | .3614 | .3543 | .3452 | .3347 | .3230 | .3106 | .2975 | .2842 | .2707 |
| 2 | .2014 | .2169 | .2303 | .2417 | .2510 | .2584 | .2640 | .2678 | .2700 | .2707 |
| 3 | .0738 | .0867 | .0998 | .1128 | .1255 | .1378 | .1496 | .1607 | .1710 | .1804 |
| 4 | .0203 | .0260 | .0324 | .0395 | .0471 | .0551 | .0636 | .0723 | .0812 | .0902 |
| 5 | .0045 | .0062 | .0084 | .0111 | .0141 | .0176 | .0216 | .0260 | .0309 | .0361 |
| 6 | .0008 | .0012 | .0018 | .0026 | .0035 | .0047 | .0061 | .0078 | .0098 | .0120 |
| 7 | .0001 | .0002 | .0003 | .0005 | .0008 | .0011 | .0015 | .0020 | .0027 | .0034 |
| 8 | .0000 | .0000 | .0001 | .0001 | .0001 | .0002 | .0003 | .0005 | .0006 | .0009 |
| 9 | .0000 | .0000 | .0000 | .0000 | .0000 | .0000 | .0001 | .0001 | .0001 | .0002 |

$\mu$

| r | 2.1 | 2.2 | 2.3 | 2.4 | 2.5 | 2.6 | 2.7 | 2.8 | 2.9 | 3.0 |
|---|---|---|---|---|---|---|---|---|---|---|
| 0 | .1225 | .1108 | .1003 | .0907 | .0821 | .0743 | .0672 | .0608 | .0550 | .0498 |
| 1 | .2527 | .2438 | .2306 | .2177 | .2052 | .1931 | .1815 | .1703 | .1596 | .1494 |
| 2 | .2700 | .2681 | .2652 | .2613 | .2565 | .2510 | .2450 | .2384 | .2314 | .2240 |
| 3 | .1890 | .1966 | .2033 | .2090 | .2138 | .2176 | .2205 | .2225 | .2237 | .2240 |
| 4 | .0992 | .1082 | .1196 | .1254 | .1336 | .1414 | .1488 | .1557 | .1662 | .1680 |
| 5 | .0417 | .0476 | .0538 | .0602 | .0668 | .0735 | .0804 | .0872 | .0940 | .1008 |
| 6 | .0146 | .0174 | .0206 | .0241 | .0278 | .0319 | .0362 | .0407 | .0455 | .0504 |
| 7 | .0044 | .0055 | .0068 | .0083 | .0099 | .0118 | .0139 | .0163 | .0188 | .0216 |
| 8 | .0011 | .0015 | .0019 | .0025 | .0031 | .0038 | .0047 | .0057 | .0068 | .0081 |
| 9 | .0003 | .0004 | .0005 | .0007 | .0009 | .0011 | .0014 | .0018 | .0022 | .0027 |
| 10 | .0001 | .0001 | .0001 | .0002 | .0002 | .0003 | .0004 | .0005 | .0006 | .0008 |
| 11 | .0000 | .0000 | .0000 | .0000 | .0000 | .0001 | .0001 | .0001 | .0002 | .0002 |
| 12 | .0000 | .0000 | .0000 | .0000 | .0000 | .0000 | .0000 | .0000 | .0000 | .0001 |

$\mu$

| r | 3.1 | 3.2 | 3.3 | 3.4 | 3.5 | 3.6 | 3.7 | 3.8 | 3.9 | 4.0 |
|---|---|---|---|---|---|---|---|---|---|---|
| 0 | .0450 | .0408 | .0369 | .0334 | .0302 | .0273 | .0247 | .0224 | .0202 | .0183 |
| 1 | .1397 | .1304 | .1217 | .1135 | .1057 | .0984 | .0915 | .0850 | .0789 | .0733 |
| 2 | .2165 | .2087 | .2008 | .1929 | .1850 | .1771 | .1692 | .1615 | .1539 | .1465 |
| 3 | .2237 | .2226 | .2209 | .2186 | .2158 | .2125 | .2087 | .2046 | .2001 | .1954 |
| 4 | .1734 | .1781 | .1823 | .1858 | .1888 | .1912 | .1931 | .1944 | .1951 | .1954 |
| 5 | .1075 | .1140 | .1203 | .1264 | .1322 | .1377 | .1429 | .1477 | .1522 | .1563 |
| 6 | .0555 | .0608 | .0662 | .0716 | .0771 | .0826 | .0881 | .0936 | .0989 | .1042 |
| 7 | .0246 | .0278 | .0312 | .0348 | .0385 | .0425 | .0466 | .0508 | .0551 | .0595 |
| 8 | .0095 | .0111 | .0129 | .0148 | .0169 | .0191 | .0215 | .0241 | .0269 | .0298 |
| 9 | .0033 | .0040 | .0047 | .0056 | .0066 | .0076 | .0089 | .0102 | .0116 | .0132 |
| 10 | .0010 | .0013 | .0016 | .0019 | .0023 | .0028 | .0033 | .0039 | .0045 | .0053 |
| 11 | .0003 | .0004 | .0005 | .0006 | .0007 | .0009 | .0011 | .0013 | .0016 | .0019 |
| 12 | .0001 | .0001 | .0001 | .0002 | .0002 | .0003 | .0003 | .0004 | .0005 | .0006 |
| 13 | .0000 | .0000 | .0000 | .0000 | .0001 | .0001 | .0001 | .0001 | .0002 | .0002 |
| 14 | .0000 | .0000 | .0000 | .0000 | .0000 | .0000 | .0000 | .0000 | .0000 | .0001 |

$\mu$

| r | 4.1 | 4.2 | 4.3 | 4.4 | 4.5 | 4.6 | 4.7 | 4.8 | 4.9 | 5.0 |
|---|---|---|---|---|---|---|---|---|---|---|
| 0 | .0166 | .0150 | .0136 | .0123 | .0111 | .0101 | .0091 | .0082 | .0074 | .0067 |
| 1 | .0679 | .0630 | .0583 | .0540 | .0500 | .0462 | .0427 | .0395 | .0365 | .0337 |
| 2 | .1393 | .1323 | .1254 | .1188 | .1125 | .1063 | .1005 | .0948 | .0894 | .0842 |
| 3 | .1904 | .1852 | .1798 | .1743 | .1687 | .1631 | .1574 | .1517 | .1460 | .1404 |
| 4 | .1951 | .1944 | .1933 | .1917 | .1898 | .1875 | .1849 | .1820 | .1789 | .1755 |
| 5 | .1600 | .1633 | .1662 | .1687 | .1708 | .1725 | .1738 | .1747 | .1753 | .1755 |
| 6 | .1093 | .1143 | .1191 | .1237 | .1281 | .1323 | .1362 | .1398 | .1432 | .1462 |
| 7 | .0640 | .0686 | .0732 | .0778 | .0824 | .0869 | .0914 | .0959 | .1002 | .1044 |
| 8 | .0328 | .0360 | .0393 | .0428 | .0463 | .0500 | .0537 | .0575 | .0614 | .0653 |
| 9 | .0150 | .0168 | .0188 | .0209 | .0232 | .0255 | .0280 | .0307 | .0334 | .0363 |
| 10 | .0061 | .0071 | .0081 | .0092 | .0104 | .0118 | .0132 | .0147 | .0164 | .0181 |
| 11 | .0023 | .0027 | .0032 | .0037 | .0043 | .0049 | .0056 | .0064 | .0073 | .0082 |
| 12 | .0008 | .0009 | .0011 | .0014 | .0016 | .0019 | .0022 | .0026 | .0030 | .0034 |
| 13 | .0002 | .0003 | .0004 | .0005 | .0006 | .0007 | .0008 | .0009 | .0011 | .0013 |
| 14 | .0001 | .0001 | .0001 | .0001 | .0002 | .0002 | .0003 | .0004 | .0004 | .0005 |
| 15 | .0000 | .0000 | .0000 | .0000 | .0001 | .0001 | .0001 | .0001 | .0001 | .0002 |

| | | | | | $\mu$ | | | | | |
|---|---|---|---|---|---|---|---|---|---|---|
| r | 5.1 | 5.2 | 5.3 | 5.4 | 5.5 | 5.6 | 5.7 | 5.8 | 5.9 | 6.0 |
| 0 | .0061 | .0055 | .0050 | .0045 | .0041 | .0037 | .0033 | .0030 | .0027 | .0025 |
| 1 | .0311 | .0287 | .0265 | .0244 | .0225 | .0207 | .0191 | .0176 | .0162 | .0149 |
| 2 | .0793 | .0746 | .0701 | .0659 | .0618 | .0580 | .0544 | .0509 | .0477 | .0446 |
| 3 | .1348 | .1293 | .1239 | .1185 | .1133 | .1082 | .1033 | .0985 | .0938 | .0892 |
| 4 | .1719 | .1681 | .1641 | .1600 | .1558 | .1515 | .1472 | .1428 | .1383 | .1339 |
| 5 | .1753 | .1748 | .1740 | .1728 | .1714 | .1697 | .1678 | .1656 | .1632 | .1606 |
| 6 | .1490 | .1515 | .1537 | .1555 | .1571 | .1584 | .1594 | .1601 | .1605 | .1606 |
| 7 | .1086 | .1125 | .1163 | .1200 | .1234 | .1267 | .1298 | .1326 | .1353 | .1377 |
| 8 | .0692 | .0731 | .0771 | .0810 | .0849 | .0887 | .0925 | .0962 | .0998 | .1033 |
| 9 | .0392 | .0423 | .0454 | .0486 | .0519 | .0552 | .0586 | .0620 | .0654 | .0668 |
| 10 | .0200 | .0220 | .0241 | .0262 | .0285 | .0309 | .0334 | .0359 | .0386 | .0413 |
| 11 | .0093 | .0104 | .0116 | .0129 | .0143 | .0157 | .0173 | .0190 | .0207 | .0225 |
| 12 | .0039 | .0045 | .0051 | .0058 | .0065 | .0073 | .0082 | .0092 | .0102 | .0113 |
| 13 | .0015 | .0018 | .0021 | .0024 | .0028 | .0032 | .0036 | .0041 | .0046 | .0052 |
| 14 | .0006 | .0007 | .0008 | .0009 | .0011 | .0013 | .0015 | .0017 | .0019 | .0022 |
| 15 | .0002 | .0002 | .0003 | .0003 | .0004 | .0005 | .0006 | .0007 | .0008 | .0009 |
| 16 | .0001 | .0001 | .0001 | .0001 | .0001 | .0002 | .0002 | .0002 | .0003 | .0003 |
| 17 | .0000 | .0000 | .0000 | .0000 | .0000 | .0001 | .0001 | .0001 | .0001 | .0001 |

| | | | | | $\mu$ | | | | | |
|---|---|---|---|---|---|---|---|---|---|---|
| r | 6.1 | 6.2 | 6.3 | 6.4 | 6.5 | 6.6 | 6.7 | 6.8 | 6.9 | 7.0 |
| 0 | .0022 | .0020 | .0018 | .0017 | .0015 | .0014 | .0012 | .0011 | .0010 | .0009 |
| 1 | .0137 | .0126 | .0116 | .0106 | .0098 | .0090 | .0082 | .0076 | .0070 | .0064 |
| 2 | .0417 | .0390 | .0364 | .0340 | .0318 | .0296 | .0276 | .0258 | .0240 | .0223 |
| 3 | .0848 | .0806 | .0765 | .0726 | .0688 | .0652 | .0617 | .0584 | .0552 | .0521 |
| 4 | .1294 | .1249 | .1205 | .1162 | .1118 | .1076 | .1034 | .0992 | .0952 | .0912 |
| 5 | .1579 | .1549 | .1519 | .1487 | .1454 | .1420 | .1385 | .1349 | .1314 | .1277 |
| 6 | .1605 | .1601 | .1595 | .1586 | .1575 | .1562 | .1546 | .1529 | .1511 | .1490 |
| 7 | .1399 | .1418 | .1435 | .1450 | .1462 | .1472 | .1480 | .1486 | .1489 | .1490 |
| 8 | .1066 | .1099 | .1130 | .1160 | .1188 | .1215 | .1240 | .1263 | .1284 | .1304 |
| 9 | .0723 | .0757 | .0791 | .0825 | .0858 | .0891 | .0923 | .0954 | .0985 | .1014 |
| 10 | .0441 | .0469 | .0498 | .0528 | .0558 | .0588 | .0618 | .0649 | .0679 | .0710 |
| 11 | .0245 | .0265 | .0285 | .0307 | .0330 | .0353 | .0377 | .0401 | .0426 | .0452 |
| 12 | .0124 | .0137 | .0150 | .0164 | .0179 | .0194 | .0210 | .0227 | .0245 | .0264 |
| 13 | .0058 | .0065 | .0073 | .0081 | .0089 | .0098 | .0108 | .0119 | .0130 | .0142 |
| 14 | .0025 | .0029 | .0033 | .0037 | .0041 | .0046 | .0052 | .0058 | .0064 | .0071 |
| 15 | .0010 | .0012 | .0014 | .0016 | .0018 | .0020 | .0023 | .0026 | .0029 | .0033 |
| 16 | .0004 | .0005 | .0005 | .0006 | .0007 | .0008 | .0010 | .0011 | .0013 | .0014 |
| 17 | .0001 | .0002 | .0002 | .0002 | .0003 | .0003 | .0004 | .0004 | .0005 | .0006 |
| 18 | .0000 | .0001 | .0001 | .0001 | .0001 | .0001 | .0001 | .0002 | .0002 | .0002 |
| 19 | .0000 | .0000 | .0000 | .0000 | .0000 | .0000 | .0000 | .0001 | .0001 | .0001 |

# Probabilities for the normal distribution

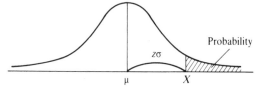

Example $Z = \dfrac{X-\mu}{\sigma}$

Probability

$P[Z>2]=0.0228$
$P[Z>1]=0.1587$

| Normal Deviate $z$ | .00 | .01 | .02 | .03 | .04 | .05 | .06 | .07 | .08 | .09 |
|---|---|---|---|---|---|---|---|---|---|---|
| 0.0 | .5000 | .4960 | .4920 | .4880 | .4840 | .4801 | .4761 | .4721 | .4681 | .4641 |
| 0.1 | .4602 | .4562 | .4522 | .4483 | .4443 | .4404 | .4364 | .4325 | .4286 | .4247 |
| 0.2 | .4207 | .4168 | .4129 | .4090 | .4052 | .4013 | .3974 | .3936 | .3897 | .3859 |
| 0.3 | .3821 | .3783 | .3745 | .3707 | .3669 | .3632 | .3594 | .3557 | .3520 | .3483 |
| 0.4 | .3446 | .3409 | .3372 | .3336 | .3300 | .3264 | .3228 | .3192 | .3156 | .3121 |
| 0.5 | .3085 | .3050 | .3015 | .2981 | .2946 | .2912 | .2877 | .2843 | .2810 | .2776 |
| 0.6 | .2743 | .2709 | .2676 | .2643 | .2611 | .2578 | .2546 | .2514 | .2483 | .2451 |
| 0.7 | .2420 | .2389 | .2358 | .2327 | .2296 | .2266 | .2236 | .2206 | .2177 | .2148 |
| 0.8 | .2119 | .2090 | .2061 | .2033 | .2005 | .1977 | .1949 | .1922 | .1894 | .1867 |
| 0.9 | .1841 | .1814 | .1788 | .1762 | .1736 | .1711 | .1685 | .1660 | .1635 | .1611 |
| 1.0 | .1587 | .1562 | .1539 | .1515 | .1492 | .1469 | .1446 | .1423 | .1401 | .1379 |
| 1.1 | .1357 | .1335 | .1314 | .1292 | .1271 | .1251 | .1230 | .1210 | .1190 | .1170 |
| 1.2 | .1151 | .1131 | .1112 | .1093 | .1075 | .1056 | .1038 | .1020 | .1003 | .0985 |
| 1.3 | .0968 | .0951 | .0934 | .0918 | .0901 | .0885 | .0869 | .0853 | .0838 | .0823 |
| 1.4 | .0808 | .0793 | .0778 | .0764 | .0749 | .0735 | .0721 | .0708 | .0694 | .0681 |
| 1.5 | .0668 | .0655 | .0643 | .0630 | .0618 | .0606 | .0594 | .0582 | .0571 | .0559 |
| 1.6 | .0548 | .0537 | .0526 | .0516 | .0505 | .0495 | .0485 | .0475 | .0465 | .0455 |
| 1.7 | .0446 | .0436 | .0427 | .0418 | .0409 | .0401 | .0392 | .0384 | .0375 | .0367 |
| 1.8 | .0359 | .0351 | .0344 | .0336 | .0329 | .0322 | .0314 | .0307 | .0301 | .0294 |
| 1.9 | .0287 | .0281 | .0274 | .0268 | .0262 | .0256 | .0250 | .0244 | .0239 | .0233 |
| 2.0 | .0228 | .0222 | .0217 | .0212 | .0207 | .0202 | .0197 | .0192 | .0188 | .0183 |
| 2.1 | .0179 | .0174 | .0170 | .0166 | .0162 | .0158 | .0154 | .0150 | .0146 | .0143 |
| 2.2 | .0139 | .0136 | .0132 | .0129 | .0125 | .0122 | .0119 | .0116 | .0113 | .0110 |
| 2.3 | .0107 | .0104 | .0102 | .0099 | .0096 | .0094 | .0091 | .0089 | .0087 | .0084 |
| 2.4 | .0082 | .0080 | .0078 | .0075 | .0073 | .0072 | .0069 | .0068 | .0066 | .0064 |
| 2.5 | .0062 | .0060 | .0059 | .0057 | .0055 | .0054 | .0052 | .0051 | .0049 | .0048 |
| 2.6 | .0047 | .0045 | .0044 | .0043 | .0041 | .0040 | .0039 | .0038 | .0037 | .0036 |
| 2.7 | .0035 | .0034 | .0033 | .0032 | .0031 | .0030 | .0029 | .0028 | .0027 | .0026 |
| 2.8 | .0026 | .0025 | .0024 | .0023 | .0023 | .0022 | .0021 | .0021 | .0020 | .0019 |
| 2.9 | .0019 | .0018 | .0018 | .0017 | .0016 | .0016 | .0015 | .0015 | .0014 | .0014 |
| 3.0 | .0013 | .0013 | .0013 | .0012 | .0012 | .0011 | .0011 | .0011 | .0010 | .0010 |

# Appendix D Critical values for the $\chi^2$ distribution

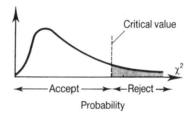

Critical value

← Accept → ← Reject →

Probability

| Degree of freedom | 0.250 | 0.100 | 0.050 | 0.025 | 0.010 | 0.005 | 0.001 |
|---|---|---|---|---|---|---|---|
| 1 | 1.32 | 2.71 | 3.84 | 5.02 | 6.63 | 7.88 | 10.8 |
| 2 | 2.77 | 4.61 | 5.99 | 7.38 | 9.21 | 10.6 | 13.8 |
| 3 | 4.11 | 6.25 | 7.81 | 9.35 | 11.3 | 12.8 | 16.3 |
| 4 | 5.39 | 7.78 | 9.49 | 11.1 | 13.3 | 14.9 | 18.5 |
| 5 | 6.63 | 9.24 | 11.1 | 12.8 | 15.1 | 16.7 | 20.5 |
| 6 | 7.84 | 10.6 | 12.6 | 14.4 | 16.8 | 18.5 | 22.5 |
| 7 | 9.04 | 12.0 | 14.1 | 16.0 | 18.5 | 20.3 | 24.3 |
| 8 | 10.2 | 13.4 | 15.5 | 17.5 | 20.3 | 22.0 | 26.1 |
| 9 | 11.4 | 14.7 | 16.9 | 19.0 | 21.7 | 23.6 | 27.9 |
| 10 | 12.5 | 16.0 | 18.3 | 20.5 | 23.2 | 25.2 | 29.6 |
| 11 | 13.7 | 17.3 | 19.7 | 21.9 | 24.7 | 26.8 | 31.3 |
| 12 | 14.8 | 18.5 | 21.0 | 23.3 | 26.2 | 28.3 | 32.9 |
| 13 | 16.0 | 19.8 | 22.4 | 24.7 | 27.7 | 29.8 | 34.5 |
| 14 | 17.1 | 21.1 | 23.7 | 26.1 | 29.1 | 31.3 | 36.1 |
| 15 | 18.2 | 22.3 | 25.0 | 27.5 | 30.6 | 32.8 | 37.7 |
| 16 | 19.4 | 23.5 | 26.3 | 28.8 | 32.0 | 34.3 | 39.3 |
| 17 | 20.5 | 24.8 | 27.6 | 30.2 | 33.4 | 35.7 | 40.8 |
| 18 | 21.6 | 26.0 | 28.9 | 31.5 | 34.8 | 37.2 | 42.3 |
| 19 | 22.7 | 27.2 | 30.1 | 32.9 | 36.2 | 38.6 | 43.8 |
| 20 | 23.8 | 28.4 | 31.4 | 34.2 | 37.6 | 40.0 | 45.3 |
| 21 | 24.9 | 29.6 | 32.7 | 35.5 | 38.9 | 41.4 | 46.8 |
| 22 | 26.0 | 30.8 | 33.9 | 36.8 | 40.3 | 42.8 | 48.3 |
| 23 | 27.1 | 32.0 | 35.2 | 38.1 | 41.6 | 44.2 | 49.7 |
| 24 | 28.2 | 33.2 | 36.4 | 39.4 | 43.0 | 45.6 | 51.2 |

| Degree of freedom | 0.250 | 0.100 | 0.050 | 0.025 | 0.010 | 0.005 | 0.001 |
|---|---|---|---|---|---|---|---|
| 25 | 29.3 | 34.4 | 37.7 | 40.6 | 44.3 | 46.9 | 52.6 |
| 26 | 30.4 | 35.6 | 38.9 | 41.9 | 45.6 | 48.3 | 54.1 |
| 27 | 31.5 | 36.7 | 40.1 | 43.2 | 47.0 | 49.6 | 55.5 |
| 28 | 32.6 | 37.9 | 41.3 | 44.5 | 48.3 | 51.0 | 56.9 |
| 29 | 33.7 | 39.1 | 42.6 | 45.7 | 49.6 | 52.3 | 58.3 |
| 30 | 34.8 | 40.3 | 43.8 | 47.0 | 50.9 | 53.7 | 59.7 |
| 40 | 45.6 | 51.8 | 55.8 | 59.3 | 63.7 | 66.8 | 73.4 |
| 50 | 56.3 | 63.2 | 67.5 | 71.4 | 76.2 | 79.5 | 86.7 |
| 60 | 67.0 | 74.4 | 79.1 | 83.3 | 88.4 | 92.0 | 99.6 |
| 70 | 77.6 | 85.5 | 90.5 | 95.0 | 100 | 104 | 112 |
| 80 | 88.1 | 96.6 | 102 | 107 | 112 | 116 | 125 |
| 90 | 98.6 | 108 | 113 | 118 | 123 | 128 | 137 |
| 100 | 109 | 118 | 124 | 130 | 136 | 140 | 149 |

# Index